T0298329

Managerial Perspective to Operational Excellence

Managerial Perspective to Operational Excellence

Using Lean Ideas to Compete Against Low-Cost Countries

Atul Tripathi, Jaymalya Deb, and Vikas Kumar

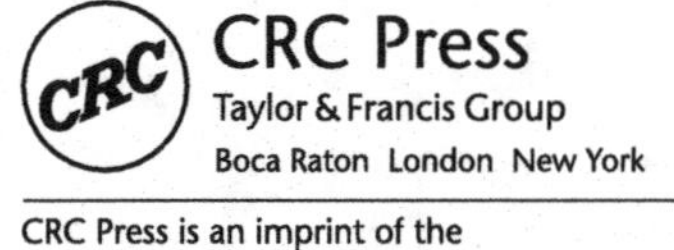

CRC Press is an imprint of the
Taylor & Francis Group, an **informa** business

First edition published 2021
by CRC Press
6000 Broken Sound Parkway NW, Suite 300, Boca Raton, FL 33487-2742

and by CRC Press
2 Park Square, Milton Park, Abingdon, Oxon, OX14 4RN

Library of Congress Cataloging-in-Publication Data
Names: Tripathi, Atul (Writer on industrial management), author. |
Deb, Jaymalya, author. | Kumar, Vikas (Writer on industrial management),
author.
Title: Managerial perspective to operational excellence : using lean ideas to compete
against low-cost countries / Atul Tripathi, Jaymalya Deb, and Vikas Kumar.
Description: First edition. | Boca Raton, FL : CRC Press, 2021. |
Includes bibliographical references and index.
Identifiers: LCCN 2020046073 (print) | LCCN 2020046074 (ebook) |
ISBN 9780367688424 (hbk) | ISBN 9781003139294 (ebk)
Subjects: LCSH: Operations research. | Lean manufacturing. | Competition,
International.
Classification: LCC T57.6 .T745 2021 (print) | LCC T57.6 (ebook) |
DDC 658.4/06—dc23
LC record available at https://lccn.loc.gov/2020046073
LC ebook record available at https://lccn.loc.gov/2020046074

ISBN: 978-0-367-68842-4 (hbk)
ISBN: 978-0-367-68843-1 (pbk)
ISBN: 978-1-003-13929-4 (ebk)

Typeset in Times
by codeMantra

Dedication

More than anything, writing a book needs significant time commitment, which in most cases means sacrificing personal time away from other responsibilities. Taking time out for this book during coronavirus-driven isolation was especially difficult since we were all trying to adjust to this new lifestyle. We therefore sincerely thank our family members for their support, sacrifices, and encouragement over the past several months. Although the list of people who have helped and guided us over the past several years is too big for this book, we would especially like to mention the following for their selfless help and guidance: Abhay Singh, Alan Yap, Alok Tripathi, Ankita Tripathi, Anup Tripathi, Prof. Andi Smart, Anju Dubey, Anushree Verma, Aparajita Deb, Asitang Mishra, Prof. Britt Fruend, Camila Pizano-Correa, Christopher Moody, Prof. Frank Pfefferkorn, Harsh Kumar, Ira Thakur, Jim Schulte, K.N. Prasad, Kavita Mishra, Prof. Manoj Kumar Tiwari, Dr. Mohit Goswami, Nirmala Tripathi, Megha Gupta, Myles Scott, Dr. Prakash, Radharaman Deb, Ragini Singh, Prof. Rajan Suri, Rajeev Thakur, Prof. Roger Maull, Sanjay Dubey, Susan Chao, Sushanta Sahu, Tebi Cheeran, V.K. Mishra, and U.C. Tripathi. This book would not be possible without the support and motivation from our parents, mentors, and advisors. Their blessings and guidance over the past several years have molded us into who we are.

Contents

Foreword

I am delighted to write the foreword for this book because I believe deeply in educating students, aspiring young professionals as well as seasoned practitioners about applying Lean and Operations Management in today's new era. Constantly changing circumstances force companies to become more agile about streamlining processes along the value chain which provide enormous leverage for any organization. Since I have been a lean practitioner and supply chain consultant myself, I found this book extremely valuable for multiple reasons, and I am sure that readers would find it too.

First of all, this book provides a case-based approach that would help managers to turn around their business when the need arises. It gradually introduces readers through the first-hand experience of steps taken to restructure three companies operating in Oil & Gas, Telecommunication, and Furniture production industries. Second, this book will serve as a practical handbook for organizations undertaking the holistic approach for any process improvement projects. It shows that understanding the problem in detail, getting buy-in from top management, and, most importantly, the context of why the change is needed are the key to the success of the project. Then it follows the top-down approach to assess the magnitude of the problem and figuring out where the problem lies before going deep into the analysis of the manufacturing process such as capacity versus utilization, lead time for particular production stage, productivity, etc. Third, despite that the focus on differentiating the company has been made on the overall lead time, the authors do a great job explaining the other differentiation strategies a company can undertake to stand out from the competition such as marketing, sales, distribution network, or unique product features to take on the low-cost country competition. Moreover, they go into further elaborating that differentiating initiatives by different functions within an organization must complement each other to reach the same destination. For example, an effort by sales to get more customers through high discounts may clash with a parallel initiative by engineering to differentiate products through customized features that would impart the uniqueness to the product.

This book is not a typical textbook or methodology book which explains all the concepts behind the Little's Law, value stream mapping, lead time, make-to-stock versus make-to-order/engineer-to-order, or reorder point. So, if something is not understood, the reader is encouraged to research on their own. From my perspective, this is a beauty of self-study, when you are not given a recipe to follow but rather figure it out on your own. I believe only by investing time in the unknown field you are able not only to retain that knowledge with higher precision but also apply learned material in practice.

The last point, but not the least, is that all three cases covered in the book touch on almost every facet of business and its impact on the business, providing readers with a broader knowledge of interconnected pieces in the operations management. Processes such as order receipt and entry, material procurement and planning, production process, and shipping are analyzed in great detail. Interestingly enough, the authors also mentioned some state-of-the-art technologies such as Auto-ML, data science, and robotic process automation that would help businesses to get some additional ideas to improve current processes.

I strongly recommend this book as a primer for final-year students, MBA students, postgraduate business and engineering students, real practitioners of lean and operations management, professional consultants, and mid-to-senior management helping them across the globe to learn, teach, and apply years of structured experience to become more resilient and competitive in a fast-paced environment.

Andrey Malenkov
Consultant, Ex-manager of KPMG, Ex-senior associate at DuPont,
Ex-head of product at Keleya

Preface

In the current age of manufacturing revival in the United States and other developed countries, companies are looking for a differentiating factor to be able to compete with competitors in low-cost countries. There are several different approaches a company could take to become more competitive depending on the life cycle of a product, product features, market, customer preferences, and industry structure. Using lean methodologies is one such operations-focused approach that has effectively helped numerous companies over the past few decades excel in what they do. However, the path to achieving a leaner organization is not always easy. There are several challenges associated in this journey including identification of project scope, use of system or tools that could help identify areas of improvement, and then implementation of new ideas to realize the benefits. Several new ideas related to Robotic Process Automation and machine learning are also gradually finding their way in a traditional manufacturing environment.

Cost-based competition is the most prevalent form of competition in mature industries/products, but the need to keep costs lower is needed even more in developed countries due to high labor and fixed costs. This book presents a case-based approach that could help leaders and middle managers at manufacturing companies structure their turnaround or improvement projects. Although the companies discussed in these case studies are based in the United States, challenges faced by these companies are the same among most companies based in Europe or the United Kingdom. The book walks the readers through first-hand experience of steps taken in actual projects with three different companies in Oil & Gas, Telecommunication, and Furniture production industries.

The details of each industry project are broken down into two chapters with the first chapter describing the problem and subsequent chapter providing details of steps taken to identify root causes and develop recommendations. The book also covers a few examples of savings resulting from the implementation of Robotic Process Automation in a manufacturing setting. We encourage readers to not only read all the cases in this book but also formulate their own strategies after reading about a company and its challenges in the first chapter of each case.

Acknowledgments

We are thankful to our alma mater, University of Wisconsin – Madison, Rice University, University of Texas – Austin, University of Exeter, National Institute of Foundry and Forge Technology – Ranchi, and R.V. College of Engineering, Bangalore, for building the foundation for our professional growth. In addition, we are thankful to our organizations and institutions (Bristol Business School, University of the West of England – UK) for providing the necessary time and support needed to complete this book on time. Lastly, we would also like to thank Cindy Renee Carelli, Erin Harris, and all the professional staff at CRC Press for their prompt support from day one.

Authors

Atul Tripathi is an operations and supply chain management professional with over 15 years of experience in engineering, manufacturing, and consulting with several Fortune 500 companies. Atul holds an MBA from Rice University, with specialization in finance and strategy, as well as dual master's in Industrial Engineering and Engineering Management from the University of Wisconsin-Madison. He received his undergraduate degree in Manufacturing Engineering from Ranchi University, India, and subsequently worked with India's largest automaker, Tata Motors, as a design engineer. In more recent years Atul has been working in energy industry with primary focus on business process improvement, asset optimization, lean manufacturing, quick response manufacturing, and business development.

Atul is also a regular reviewer for many high impact international journals and conferences and has been a guest speaker on operations and change management at several APICS conferences. His areas of interest include operations management, change management, corporate finance, and strategy.

Atul lives with his wife, Urvashi, and son, Veer, in Houston, TX, USA. Outside of work he is an avid traveler, a news junkie, and a fitness enthusiast. Atul also volunteers as ESL tutor for non-English speaking immigrants.

Jaymalya Deb is a supply chain professional. A graduate with distinction in Mechanical Engineering from RVCE, India, Jay also holds a master's degree in Manufacturing Systems Engineering from the University of Wisconsin-Madison and an MBA from The University of Texas at Austin. Jay started his career working for Toyota and was introduced to the Toyota Production System, which built his interest in lean manufacturing. After this, Jay has worked in multiple roles in manufacturing operations, supply chain management, and ERP systems implementation for over 15 years with Fortune 500 companies.

Jay's primary area of interest is in supply chain optimization, especially as it relates to demand forecasting to assist businesses to run a lean, scalable supply chain that keeps costs low and profits high. Jay currently works with researchers and consultants in this area to help develop customized machine learning and deep learning models for demand forecasting.

Jay lives with his wife, Nidhi, and daughter, Aaradhya, in Fort Worth, TX. Outside of work, Jay enjoys traveling, yoga, volunteering, and working out.

Vikas Kumar is a Professor of Operations and Supply Chain Management and Director of Research at Bristol Business School, University of the West of England, UK. He holds a PhD degree in Management Studies from Exeter Business School, University of the Exeter, UK, and a Bachelor of Technology (first class distinction) degree in Metallurgy and Material engineering from National Institute of Foundry and Forge Technology (NIFFT), Ranchi University, India.

He has over 15 years of experience working in the area of operations and supply chain excellence. He has published more than 200 peer-reviewed articles in leading international journals and International conferences including the *Journal of Business Research, International Journal of Production Research, Supply Chain Management: An International Journal, Expert System with Applications, International Journal of Production Economics, Computers & Industrial Engineering, and Production Planning & Control.* He has also co-authored three books on quality management systems, innovation, operations, and sustainability. He has guest-edited several special issues in high impact peer-reviewed international journals. Prof. Kumar serves on the editorial board of several international journals and is a regular reviewer for many high impact international journals and conferences. He is also regularly invited as keynote speaker at major international conferences. Prof. Kumar has successfully secured funding in the excess of £1 million from various research agencies including EPSRC, Innovate UK, British Academy, British Council, and Newton Fund. He has worked on several international projects with Vietnam, Thailand, Indonesia, Turkey, Brazil, Mexico, Taiwan, and EU. His current research interests include Operational Excellence, Industry 4.0, Sustainability, Circular Economy, and Food Supply Chains.

Vikas lives with his wife, Archana, and son, Viaan, in Bristol, UK. Outside of work, Vikas is an avid traveler and enjoys watching movies, eating out, swimming, and badminton.

1 Introduction

Challenges to Change Initiative: History and Application of Lean Concepts

The idea of entering new markets like India and China in late 20th century and then later realizing lower costs of manufacturing by making products in large volumes in countries like China, Vietnam, India, and Mexico fundamentally transformed the global manufacturing landscape. The economies of scale combined with lower operating costs in these countries not only allowed companies to increase their margins, but it also improved the overall living standards of consumers all over the world. However, it also led to a mass departure of core manufacturing activities from developed countries to low-cost countries. The original model of outsourcing manufacturing to low-cost countries has also evolved over the last couple of decades due to advancements in lean supply chain philosophies. The optimization of the global supply chain to minimize inventory along the value chain and continued pressure to keep resource utilization high have made it even more susceptible to failure in an interconnected world.

However, over the last few years, there has been an increased focus toward manufacturing revival in the United States and other developed countries. The established model of outsourcing has been challenged by many long-term and short-term factors. These include changing customer preferences toward customized solutions with minimal wait time, recent geopolitical changes as well as global pandemic like COVID-19. Many companies in the developed world are looking at ways to differentiate from the competition and be better positioned to compete with players from low-cost countries. Lean manufacturing is one such approach that could help companies achieve that objective. Over the last several decades lean manufacturing has been enriched by both results from implementation and innovations in industries as well as by research contributions from universities. However, the path to achieving a leaner organization is unique for each organization and that is not always easy.

The history of lean manufacturing is more than two centuries old, from the concept of using interchangeable parts in 1799 to standardization of work concept introduced in the Ford production system in 1913. The Ford production system combined the concept of interchangeable parts with the flow of production and standardization of work. The Ford's approach lacked the flexibility to adjust to variations in product model or design, and this problem was ultimately addressed by Toyota after World

War II through a combination of different concepts that are now commonly known as Toyota Production System. The Just-In-Time (JIT) system used by Toyota aimed at increasing efficiency by receiving parts only when they are needed at each step of the manufacturing process. This was used in conjunction with a relentless focus on mistake proofing so that quality issues combined with a lack of inventory buffer do not impact on-time delivery performance. These two core principles of JIT and mistake proofing subsequently evolved into a pull or Kanban system. The Toyota Production System introduced many concepts that are now established practices among most manufacturing companies around the world. These concepts include cellular manufacturing, Single Minute Die Exchange (SMED), defining waste and focus on its elimination, Kaizen or continuous improvement aimed at reducing waste, Poka-yoke or mistake proofing, and Kanban or pull system.

Over the last several decades, lean concepts have proliferated well within education institutions and manufacturing companies, so the new batch of managers leading the charge of process improvement in different companies also have some background in lean concept. The Lean approach highlights eliminating or reducing seven types of waste namely defects, overproduction, waiting, transportation, inventory, motion, and over-processing. Due to its origin in high-volume environment of Toyota Production System combined with the lean education background of recent managers, this approach is relatively easier to get acceptance on and implement in companies that work in high-volume, low-mix environment. Over the years, our accounting practices have also evolved to support many of the lean concepts, but the financial measurements in practice may not always align perfectly with lean philosophies. For example, measures such as unit cost or gross margin on costing side or using machine utilization and efficiency in operational expense planning side are established metrics and have become part of lingo among managers or leaders in most companies.

If these measures become the primary drivers of performance, they could be detrimental to inventory positions or to acceptance of cellular manufacturing. Let us try to elaborate a bit more on this topic. A piece price-driven approach incentivizes materials managers or buyers to either make bulk purchases to get a better discount or prefer cheaper offshore vendors at the expense of delivery lead time. Bulk purchases could lead to excess on-hand inventory resulting in cash tied to the working capital of a company. On the other hand, a longer lead time necessitates planning well ahead of time based on the forecast as well as keeping safety stocks to mitigate the risk of stockouts. The management effort of long-winded supply chains with long lead times mitigates any immediate efficiencies gained through reduction of costs. Traditional or established costing systems do not include inventory holding cost as a part of the material cost or take into consideration the total landed cost of the part. The inventory holding cost includes opportunity cost in the form of cost of capital, cost of inventory management or upkeep, property tax, and the cost of obsolescence in case of design changes. For most companies, the cost of capital makes up majority of the holding cost associated with inventory.

Managers are expected to manage costs and protect throughput. These are often at cross purposes with each other. The metric to keep machines highly utilized could incentivize shop managers to make decisions that are not always aligned with lean objectives. One way to keep machines highly utilized is to run large lot sizes so theirs is minimal setup involved from part to part. In instances where the company has to outsource machining or fabrication work due to capacity limitations, a utilization-based mindset could incentivize shop manager to outsource higher margin but complicated manufacturing or fabrication processes and keep simpler and higher volume, but much lower margin parts in-house. Making such a decision would undermine the core competency of the manufacturing organization. Furthermore, a cellular manufacturing based approach requires collocating machines to process a group of parts with similar processing requirement to provide agility to the manufacturing process. Even though efforts are made during the design of a manufacturing cell to balance the load well across different machines, a utilization-based decision making approach will incentivize managers to lean more toward a function or process-based manufacturing layout than cellular layout.

Most finance or accounting managers are trained in traditional accounting practices and metrics which combined with previously established processes at a firm make it easier to implement or continue traditional performance metrics. In a high-volume and low-mix environment, these metrics are relevant as well as easily implementable. However, changing consumer preferences, increased competition, and better integrated supply chains over the last few decades have resulted in an increase in variety of products that any company has to keep in its portfolio. This transition toward a high mix and lower volume has forced most companies to move from a make-to-stock to make-to-order or engineer-to-order environment.

In this changing operational environment, quite a few new philosophies have evolved that incorporate the basics of Just-In-Time and lean concepts, but approach continuous improvement with their own core focus. One such philosophy is Quick Response Manufacturing (QRM) that focuses on relentless reduction of lead time, which ultimately leads to reduction of all other types of wastes. Competing based on lead time is a powerful concept that not only leads to a leaner organization but also allows companies in developed countries to position themselves differently to compete with players in low-cost countries. QRM is based on the principles of time-based competition and it can be implemented in all stages of value addition from shop floor to office to suppliers.

In practice, however, most companies use concepts covered by a combination of different approaches in a way that is custom to their business needs. This happens partly because of the history of a company. Changes in business model and leadership result in subsequent tilt of the organization toward a new approach of doing things. Older the organization, longer is the list of changes implemented. Without a distinct vision or focus for the process improvement at the highest level, different divisions or departments focus on initiatives that are within their control and are endorsed by the departmental managers. When a new operations leader takes over in such a scenario, the task of implementing a companywide change is not easy.

There are several implications of moving forward with a change journey because the success depends on a combination of recognizable benefit along with the acceptance of change by employees impacted. The first project needs to be significantly big to make a positive and recognizable impact on company's performance, but at the same time there are risks associated with making the scope too big to succeed. The process of carving out a sizeable, but well-contained project focus should therefore get the attention it deserves.

Initiating a transition also comes with the challenge of an effective change management process throughout the journey. Concerns should be addressed in an effective manner and success must be celebrated. One of the challenges business managers face is related to reorganizing responsibilities such that a smaller number of people take a wider set of responsibilities. For example, in many companies planning and purchasing are two different roles and require slightly different skillsets. Planning is an internal operation focused activity and requires skills such as good product knowledge, familiarity with manufacturing processes, and a good understanding of company's ERP system. On the other hand, purchasing is an externally focused activity and requires skillsets such as good financial acumen, negotiations, interpersonal skill, industry, and market awareness. Material managers prefer to have these two different roles to mitigate the impact on business activities due to employee turnover. In case of a unionized workforce, there are additional challenges associated with negotiating newly defined roles that overlap with two or more existing roles. However, a manufacturing cell works best when its members take complete ownership of its performance and work as a team. A transition toward an ownership-based cell team will not only require moving away from traditionally defined roles but will also need new incentives and reward structure.

Ultimately, each change journey is a project and it may require investment in some cases. Every project needs management of the bottleneck and monitoring the critical chain of activities. Monitoring the bottleneck as all projects come into play is key to protecting the overall throughput of the organization. In some cases, there might be more than just one bottleneck. Identifying those areas and appropriately buffering those resources will achieve the intended result. When an investment decision for a project needs to be evaluated, sometimes the traditional payback period and net present value do not take into consideration the most important factor – "availability of money." The answer on the two-dimensional problem of time and money is not an easy one to intuitively understand, since our metrics have been skewed one way or another throughout our entire work lives. Perhaps the simplest explanation we have encountered so far is –"Money is measured in dollars, investment in dollar-days" – Eli Goldratt, *Critical Chain.*

As a leader tries to move forward with his or her own vision for the company, it helps to have a consistent goal of operational excellence for the entire organization driven by a primary performance metric. This book is a collection of three unique projects that were initiated by leaders at these companies to bring about the change. This book aims to present a case-based approach that could help leaders and middle managers at manufacturing companies structure their turnaround or improvement projects. This book walks the readers through the firsthand experience of steps taken in actual improvement projects with three different companies in Oil and Gas, Electronics, and Furniture Production. We hope that the lessons learned from these case studies will benefit managers in manufacturing organizations to have a better understanding of the benefits and challenges of lean approach that will further assist them in making key strategic decisions to maintain their competitiveness.

2 Overview of Cases

This chapter provides an overall summary of all three cases discussed and is intended for readers who would like to get a quick overview of the book. We encourage readers to skip this chapter if they would like to work through the case on their own. Each of the three cases discussed in this book is broken down into two chapters, with the first chapter describing the business problem and the second chapter covering the analysis performed to identify areas of improvement. By reading the first half of each case, readers could try to formulate their own approach toward solving the problem and then read through the second half to compare it with actual analysis performed. Even for the readers who read through the summary in this chapter, the detailed chapters provide a much deeper insight into problem solving approach used by the teams working on those projects.

The first case discusses a contract job shop company, American Pride Manufacturing (Fictional name), located in San Angelo, TX. The company has been a supplier of sheet metal fabricated and stamped parts to the Oil and Gas industry since 1948. They had a union facility with a workforce of 45 employees, out of which 30 are shop floor operators. American Pride Manufacturing had annual sales of $7M at the time of the project.

The parts were produced using either stamping or fabrication processes primarily using carbon steel sheets. Almost 40% of total active part numbers used fabrication but constituted only 4% of the total production volume. Approximately 90% of all the fabricated or stamped parts required painting or plating operations, which were outsourced to a local vendor. The fabrication used laser cutting machine, brake press, and CNC punch while the stamping operations used punch presses. The customer orders were scheduled using an 8-week forecast and a 2-week hot list (firm orders) provided by the OEM. The 8-week forecast was also used in ordering the raw material.

American Pride Manufacturing was facing increased competition from companies located in low-cost countries who were able to supply parts at a lower price. However, they required large order volumes and quoted lead times of 10–12 weeks. The company owner, Joseph Schultz, believed that they could compete with these companies as well as increase their customer base by offering products in less than a week and in smaller volumes. He estimated a 25% increase in sales volume if the lead time for fabricated parts were reduced to 1 week. Joseph also planned to use lessons learnt from this first project as a reference for making changes in other areas of the company.

Joseph built a core team from manufacturing, maintenance, and purchasing to identify and implement improvement opportunities. After mapping the lead time for the facility, the team observed that the total lead time of 36 days for fabricated parts consisted of 21 days of raw material supply, 7 days of queue time for raw materials, and 8 days of fabrication. It was also observed that the actual touch time was less than 6% of the total lead time. By comparing the value-added time with the total lead time and identifying potential improvement opportunities, the team decided to reduce the manufacturing lead time for parts produced in the fabrication cell by at least 50% from an average of 36 days to 18 days. The scope of the project was set from the time an order is placed for raw materials to the time finished goods are shipped to the customer.

The team performed analysis in three different areas namely finished goods inventory (FGI), fabrication cell, and raw material supply and the following key insights were obtained:

- The frequent changes in design and the company's policy of running large batches to minimize setups had resulted in a high level of FGI. The old designs become service parts which are ordered infrequently. The FGI was being maintained for 361 parts out of a total of 600. However, more than 200 out of 361 were service parts.
- The punching and welding equipment were located far from the rest of the fabrication equipment resulting in a significant level of material handling. To minimize this, the existing operating practice involved running large batches, which in turn resulted in longer lead time.
- Due to the long lead time of 4 weeks associated with the raw material procurement, they were ordered in quantities of 5000–10,000 lbs. The raw material then waited until the previous batch was used up. The practice of running large lot sizes also sometimes resulted in the raw material being used up before the next batch was received resulting in longer response time.

Though data analysis and manufacturing process modeling in a simulation software, following key recommendations were developed:

- Create a fabrication cell by dedicating the 4000 W laser to fabricated parts and by moving spot welding, MIG welding, and 35T punch press into the fabrication cell.
- Implement a make-to-order strategy for the fabricated parts. Use smaller lot sizes for the fabricated parts and implement transfer batching.

Further discussions were carried out with the existing sheet steel supplier for 14 grades of steel which represented nearly 76% of the total raw material requirement. From discussions with the sheet steel supplier and American Pride Manufacturing personnel and analysis of raw material usage data, the following key recommendations were made for the raw material procurement:

- Use the 8-week forecast to estimate raw material requirement and order raw material on a weekly basis for the identified grades of steel.
- Update the raw material distributor about current Raw Material Inventory (RMI) status and the requirements for an 8-week period.

A 2-month implementation plan was developed, which included shop floor rearrangement, raw material supply management, and operator training. The implementation process was initiated with equipment relocation and operator cross-training. This was followed by changes in the lot-sizing policy and raw material procurement policy. The support of the management for the program and training of employees on the benefits of lead time reduction helped create the momentum for change.

The implementation of these recommendations required a small initial investment amounting to a total one-time cost of $5K for the equipment installation and operator training. The reduction in RMI, FGI, and Work in Process (WIP) was estimated to result in a one-time saving of approximately $34K. Additionally, an annual increase in profit of approximately $40K was expected from additional sales.

An implementation team was also created which consisted of the quality manager and key personnel from the union who were interested in this change process. The implementation of these recommendations not only created a better business opportunity for American Pride Manufacturing but also helped weave the philosophy of continuous improvement into the cultural fabric of the company.

The second case discusses an enclosure solutions and custom contract manufacturing company, Custom Cabinet Builders (Fictional name), located in a 500,000 square foot facility in Boulder, Colorado. The company designed, manufactured, and assembled a wide range of standard and configurable cabinets, enclosures, frames, racks, and mounting products used by data centers and telecom companies worldwide. These products are sold directly to OEMs, integrators, and end-users.

Custom Cabinet Builders (CCB) built both standard and custom products based on customer needs. The standard products include preconfigured cabinets, consoles, frames, and rack mounts with options for different sizes. Custom products on the other hand are made to order according to the customer specifications. Both standard and custom products mainly use sheet steel and aluminum or steel plates as raw material. The fabrication of these parts was carried out in two distinct functional areas using Numerically Controlled (NC) punch presses, brake presses, and manual and automatic welding stations for shearing, punching, bending, and welding of parts. The fabricated parts are then painted and assembled before shipping to the customer.

Prior to 2001, CCB was the market leader in its industry, but had since struggled to be competitive and has fallen between fifth to seventh out of 25 companies in their market. The entry of similar products from low-cost countries further challenged CCB's position in the market. Most of CCB's top competitors in the United States competed by stocking inventory in warehouses close to customers and had effectively reduced their external lead time for standard products to one day. CCB did not have the resources to match their competitors' strategy. Alex Stolz was hired as the Operations Director to turn the facility around and regain its lost market position. Alex strategized adopting a lead time-focused improvement approach that would give them the opportunity to become competitive and responsive while reducing inventory and other costs. Through analysis of sales data and discussion with CCB staff, two product types belonging to the standard product family were identified as project focus. These products were Printed Circuit Board Enclosures (PCBE) and Cellular Tower Cable Racks (CTCR). The company's sales team expected that with a reduced lead time for these products, CCB Products could expect sales revenue to increase over 15% year over year.

By comparing the value-added time with the current lead time and identifying the improvement opportunities, the team decided to aim for reducing lead time by 80% for the two selected products.

Alex and his project team analyzed information received from meetings and discussions, MRP data, part routings, and supplier feedback. Five problem areas were identified: customer order flow, CTCR purchased/raw material flow, PCBE purchased/raw material flow, CTCR fabrication flow, PCBE fabrication flow and paint. Following key insights were obtained after a detailed study of the problem areas:

- Thirty percent of all the orders received by order entry department contained incomplete or inaccurate information and the order entry staff had to call back customers to get complete details.
- High variability in the number of orders received each day resulted in a backlog in order processing.
- Raw material supply for PCBE and CTCR had a lead time of 5–7 weeks which not only caused CCB to forecast their sales for 6 weeks but also to maintain a safety stock of raw material to account for forecast uncertainties.
- Some of the operations in the fabrication area were performed on shared resources resulting in large lot-sizing policy to minimize the number of setups.
- To minimize the number of setups for color change in paint line, CCB considered a time window of 5 days and all the parts of a specific color that required shipping within that 5-day period are painted in a single lot.

The following key recommendations were provided to help reduce lead time for the products:

- Develop standard order reference sheet to minimize the number of incomplete orders.
- Develop a web-based product configurator for ordering items available directly on the CCB website. Implementation of this product configurator would resolve issues related to conflicting information on incoming sales orders, reducing rework time needed with customers.
- Implement Robotic Process Automation in order for the entry department to automate entry of regular orders and slowly scale it up to manage more complex orders that would involve interaction with sales personnel, possibly using a chatbot.
- Implement Machine Learning techniques for forecasting and periodically tune it for accuracy to reduce on-hand inventory.
- Procure cut-to-size PCBE side face and PCBE base face from Colorado Extrusion, Co., which has a lead time of 3 weeks.
- Procure standard length CTCR side and cross tubes and PCBE top face from distributors, which has a lead time of 2–3 days. In-house sizing of these parts would require purchase of one new CNC band saw and move the existing bandsaw to the proposed cell.

- Create a collocated cell containing cross-trained team members for PCBE and CTCR fabrication.
- Paint parts that were going to be shipped next day.

The implementation process was planned to start with the creation of an "office cell" followed by creation of a "manufacturing cell." Creation of the "manufacturing cell" started with the purchase of a bandsaw and completion of material testing of standard materials procured from the distributor. The next step was clearing the floor space for implementation of the proposed cellular layout and subsequent cross-training of cell team members after working out work responsibility details with the union. The project team also suggested starting work on the supply chain to continue looking for lower cost and shorter lead time players in the market. Both Alex and the team believed it'll be best for the CCB to undertake a phased approach to reduce the paint MRP lead time to 1 day. It would also be important to support this reduction by resolving the identified obstacles. Additional time reduction in the paint area could be achieved through investing in an infrared oven, but this option was reserved for future improvements.

Cost justification was worked out for the project for a period of 5 years considering 15% increase in year over year sales. One-time costs associated with the implementation plan totaled $217,860 in the current year, which included new equipment purchase, equipment relocation cost, and tooling costs for in-house sizing of raw materials. The payback for the project was calculated to be 2.3 years.

It was expected that a few challenges would surface during the implementation of proposed recommendations. However, Alex envisioned they could be resolved with training, working with the labor union, and management support. Lead time reduction for PCBE and CTCR would provide Alex and his project team to make incremental changes in other product lines as well. Ultimately, it would allow CCB Products Corporation to gain its leadership position in their market through increased sales and profits.

The third case discusses Ubercraft Cabinets (Fictional name), which is located in a 200,000 square foot facility in Jasper, Indiana. They were a leader in the design, manufacture, and assembly of a wide range of cabinets, consoles, shelves, and other furniture for kitchen, bath, and other rooms. The facility had full, up-to-date veneer, manufacturing, finishing, and installation capabilities. The products built by Ubercraft were sold through distribution and large retail stores such as Home Depot and Lowes.

The products manufactured at Ubercraft could be classified into two broad categories: standard and custom. The standard products, which were called "branded products" in Ubercraft, included standard-sized cabinets, consoles, and shelves with options for different sizes. Custom products, which were called "contract manufacturing products", on the other hand, were made to order according to the customer specifications. Both standard and custom products mainly used wood, glass, and steel plates as raw material. The processes used at Ubercraft for manufacturing products included cutting, bending, painting, laminating, and assembly of such materials. The production of these parts was done using NC saws, NC drills, and miter edging machines. The finished parts were then painted and assembled before shipping to the customer. Presently, contract manufacturing sales equate to $30 million, which is 75% of the total company revenue, while the branded products equate to the remaining 25%.

Justin Sherman, the Vice President of Operations at Ubercraft, was very keen on making several changes at the company to better position it to meet customer expectations and retain its market share. Ubercraft was in the middle of a companywide lean implementation and had previously implemented changes in its high-volume segment of branded products area. To expand the concept further, Justin wanted to have another project in a different area of the company with the intent that resulting changes would be making a significant impact. With contract manufacturing being the bigger part of the business, it was the obvious choice to look for a manufacturing process improvement initiative. The lead time for contract manufacturing products was long by industry standards, especially for consistent orders. Some of Ubercraft's customers used multiple-sourcing strategy, creating additional pressure on Ubercraft to perform better than the competition. Justin believed that competing on delivery time could help Ubercraft strengthen its relationship with customers and hopefully get a larger share of orders in future. The business in the contract manufacturing area consisted of three major groups: group 100, 200, and 300 with shares of 26%, 30%, and 44%, respectively. Even though group 100 was the smallest in terms of business, its share had increased steadily in the past 3 years. In addition, Ubercraft's sales team believed its long lead time of 6 weeks was a challenge in getting more business. Within this group, Custom Miller Cabinet (CMC) was selected as the focus for this project due to its highest sales and shorter delivery expectation. To analyze and make changes in the contract manufacturing area, Justin called in the same team that had worked on the previous project for standard products.

Through initial discussions and subsequent verification through data collection, the team realized that the total time taken for production of one lot of 18 pieces was 6 weeks, with only 2 weeks being used for actual production, while the rest was waiting time. By comparing the value-added time with overall lead time and identifying potential improvement opportunities, the team set its goal to reduce the internal lead time of CMCs by 50% from 6 weeks to 3 weeks. The scope of the project included raw material supply and shop floor operations, including saw cutting, laminating, and finishing operations.

The team also conducted a brainstorming session with relevant leads and identified four main problems which made them inefficient: (1) frequent order changes in order entry, (2) shop floor capacity constraint, (3) raw material supply management, and (4) poor quality of the painting. Justin had already hired an external company to resolve paint problems, so paint operations were excluded from the project scope.

The team obtained several key insights after a detailed study of the three problem areas which are summarized below:

- Long lead times of 6 weeks made the forecast highly inaccurate so that nearly 40% of ship dates were rescheduled by request of customers, which in turn made the lead time even worse.
- Frequent order changes and use of inaccurate lead time estimates in MRP led to continuously increasing backlog.

- Insufficient labor capacity at edging and drill stations resulted in bottlenecks.
- Reordering points for supplier managed raw material inventory, which is also shared by standard products, is calculated incorrectly. High variability in demand of standard products led to occasional shortage of raw materials.
- Purchase-to-order strategy used for inventory not managed by supplier and inaccurate estimation of production lead time in the MRP system led to occasional shortage of raw materials.

By modeling the shop floor process and analyzing alternatives combined with discussions, the team developed the following set of key recommendations to realize improvements:

- Reassign operators to increase available capacity of constrained resources.
- Change production lot size to avoid grouping multiple orders and to reduce lead time.
- Change procurement strategy of non-vendor managed inventory (VMI) parts to achieve shorter lead time.
- Take a holistic approach at planning demand for vendor managed raw material inventory.
- Explain the benefits of Ubercraft's shorter lead time on CMCs to customers to reduce future order changes and encourage ordering for smaller batches based on actual demand.
- Establish a new process for Master Data Management in ERP system.
- Implement Robotic Process Automation to automate manufacturing scheduling.

The team also proposed a ten-step implementation plan that covered all the recommendations in three areas including shop floor rearrangement, raw material supply management, and order placement strategy. The implementation was suggested to begin with layout changes and operator cross-training and reallocation. This was followed by recommended lot-size changes, correction of lead time values in the MRP system, and changes in the raw material management policy.

The team expected their recommendations would help Ubercraft benefit from both one-time and annual cost savings. In addition the recommendations will also reduce the lead time for CMCs by 55% from 29 days to 13 days. The reduction in WIP of CMCs by 67% was estimated to help Ubercraft realize a one-time saving of nearly $6300. The sales group at Ubercraft indicated that with a reduced lead time they could increase the overall sales target for CMCs by 5% next year from a combination of increased order from existing customers as well as through acquisition of new customers. The overall project recommendations were expected to have a payback of less than a year.

3 Competing with Low-Cost Countries

A Steel Fabrication Shop in Texas

American Pride Manufacturing is an ISO 9002 certified contract job shop company located in San Angelo, TX. The company has been a supplier of sheet metal fabricated/stamped parts to the Oil and Gas industry since 1948. At the time of this project, American Pride Manufacturing had a workforce of 45 employees out of which 10 are office staff and 30 are shop floor operators. The company ran two shifts a day with 20 employees in the first shift and 10 in second. The company had annual sales of $15M. The products manufactured at American Pride Manufacturing were primarily sold to companies in the Oil and Gas industry. Key customers included Jack Drilling Services (JDS) Company and American Oil Services (AOS) Company which, respectively, account for 80% and 13% of the total volume.

The products used sheet steel or tubes as raw material and were primarily used in the Oil and Gas industry. These products were produced using either stamping or fabrication processes. The decision on using stamping or fabrication operations depended on part design and volume. The stamping operation was preferred for high-volume products with simpler designs. Nearly 40% of the total active part numbers used fabrication, but the volume of fabricated parts was much smaller than that of stamped parts. Figure 3.1 shows the comparison between the fabricated and stamped parts in terms of the number of active part numbers and volumes.

The sheet metal thickness used for most parts varied from 13 to 4 gauge. Carbon steel sheets accounted for 80% of all the raw material and were procured through different distributors. However, 80% of these came from two suppliers located in Houston, TX. Sheet steel was procured in standard lengths of 4′ × 7′ for the fabricated parts. However, for most of the repeat orders in the stamping area, the sheet steel was procured cut to size. The fasteners for JDS parts were procured from a fastener supplier in Sweden and some local suppliers. The raw material lead time for most grades of sheet steels was 4 weeks and up to 12 weeks for highly polished stainless steel sheets.

The customers shared an 8-week forecast for all the parts, which included a 10-day hotlist representing firm orders. For a repeat order, the parts were scheduled for production according to the ship dates. The repeat orders accounted for 85% of total production and were scheduled directly into the system by the shop floor manager who had access to both forecasts and firm orders. In addition to the firm orders, the shop floor manager also used the 40-day forecasts to determine the actual production

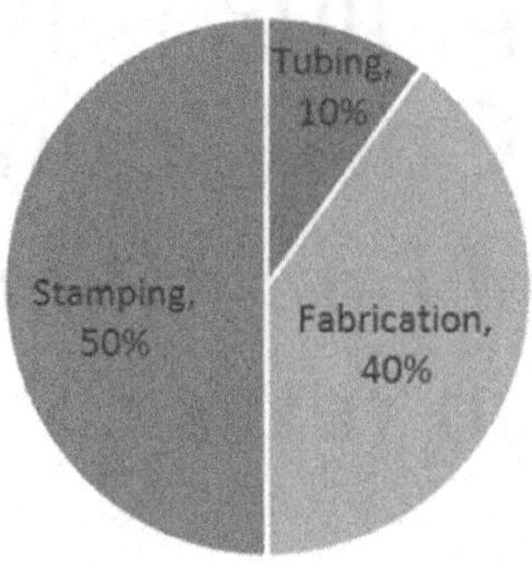

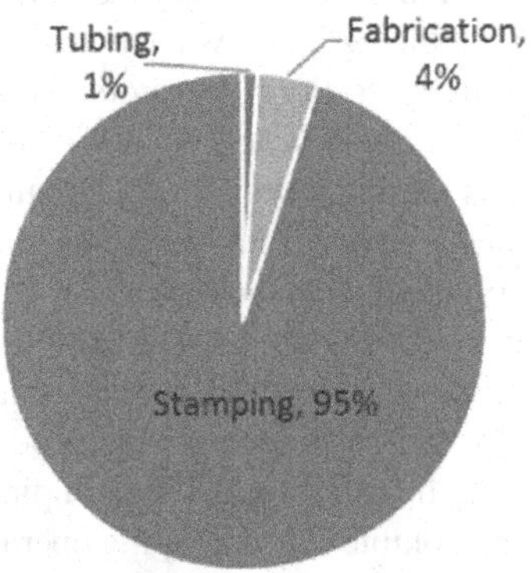

FIGURE 3.1 Fabricated versus stamped parts.

quantities. The 40-day forecast was also used in ordering the raw material. For a new order, the part drawing provided by the ordering company was used by the shop floor manager to determine the process requirements for that part. For a stamped part, this involved getting quotes from the tooling suppliers located in Midland, TX. As mentioned earlier, this decision on stamping or fabrication was dependent on the volume and complexity of the part.

The shop floor operations were carried out in two specific areas, namely fabrication and stamping. Nearly 90% of all the fabricated or stamped parts require painting or plating, which was outsourced to Prime Plating Inc., located in Midland, TX. Figure 3.2 shows the high-level process map for both stamped and fabricated products.

The stamping area had a functional layout while the fabrication area at the time was being reorganized as a cell. The fabrication area used laser cutting machines, brake press, and CNC punch, while the stamping operations used punch presses. The assembly of parts in both the fabrication and stamping areas was done through welding, riveting, or using other fasteners like cage nuts. All the welding and assembly operations were performed at a single location on the shop floor.

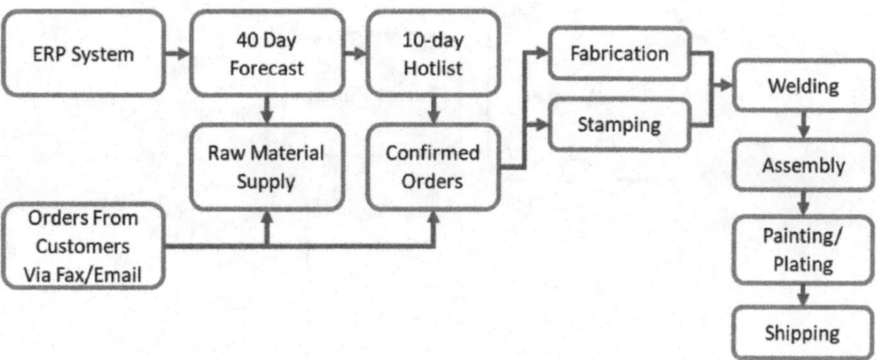

FIGURE 3.2 High-level process map.

All the equipment for stamping and fabrication processes was housed in one of the two buildings namely the north and south building. The fabrication process started from the laser cutting machine or CNC punch, which were in the south building. The parts would then go through part number stamping operation on a 35T Open Back Inclinable (OBI) press, which was in the north building. After getting the part number punched, the parts were brought back to the south building for operation on a press brake. The formed parts are again sent to the north building for welding operation. The welding workstations were shared across all the products in the company. The welded parts were finally sent for painting or plating operation to an outside supplier. Figure 3.3 summarizes the flow for the fabricated parts across the two buildings.

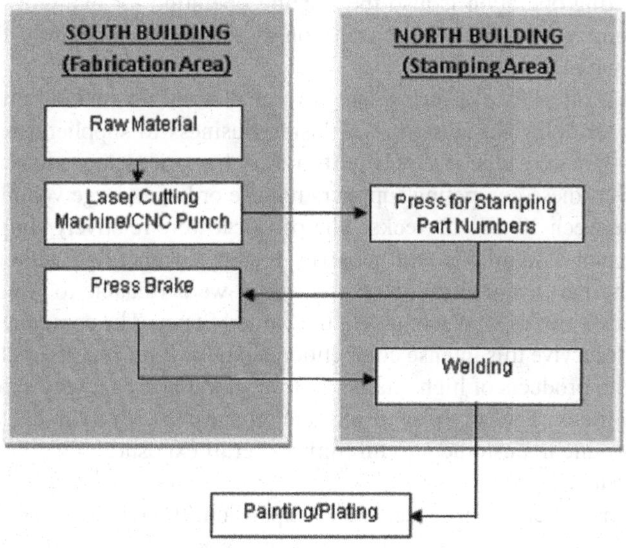

FIGURE 3.3 Process flow for the fabricated parts.

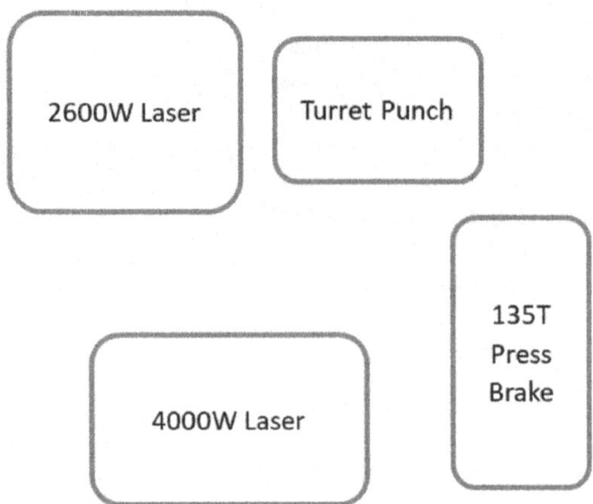

FIGURE 3.4 Layout of the fabrication area.

The existing fabrication area currently housed two laser machines of different capacity, a CNC punch, and a press brake. Figure 3.4 shows the layout of the fabrication area. The two laser machines ran two shifts with one operator allocated to them in each shift while the CNC punch and press brake ran one shift with one operator allocated to each. The raw material for both stamping and fabrication areas was stored in the south building.

The two laser machines had different capacity in terms of the thickness of material that they could cut. The laser cutting machine with higher capacity was recently purchased and the migration of the programs from the old laser machine to the new laser machine was not complete. As a result, the routing of parts to one of the two machines was based on available capacity, as well as the availability of part program. The laser cutting operation is also the starting operation for many of the stamped parts and therefore the available capacity on these two machines was also used for the stamped parts.

American Pride Manufacturing had several reasons for making changes in its existing way of doing business. It was losing business to suppliers from low-cost countries, as they were able to supply parts at very low prices due to lower production costs. However, these competing suppliers require orders in large volume and quote lead times between 10 and 12 weeks. The products were relatively simple to make, so there was not a significant entry barrier for any competition, although quality, reliability, and part availability on shorter notice were valuable to American Pride Manufacturing's two main customers in the United States. The company believed its best option to survive this intense competition from low-cost players was to differentiate itself with products of high quality that were available in a very short lead time of less than a week. The company also wanted to have the ability to deliver products in smaller volume to customers to minimize overall exposure to working capital in the value chain.

Furthermore, after almost a decade of expansion, the Oil and Gas industry had seen a significant reduction in business activity from 2015 onwards. This was primarily driven by a drop in oil prices in late 2014. Just like 2015, 2016 was also predicted to be a low demand period because of slowed drilling activity in the United States. Because of this estimated decrease in the market demand and a new supplier development policy, JDS had planned to discontinue 800 small suppliers and work with a smaller number of key suppliers. American Pride Manufacturing wanted to increase its capacity to produce more product variety along with a short lead time to strengthen its association with JDS.

A breakdown of American Pride Manufacturing's total production volume shown in Figure 3.5 revealed that JDS and AOS accounted for nearly 93% of the total production volume. This made the business at American Pride Manufacturing heavily dependent on the performance of these companies and the oil and gas drilling activity in general. The company president Schultz believed that the impact of changes in the drilling industry could be reduced if American Pride Manufacturing expands its customer base to other industries. A shorter lead time and an increased capacity would help American Pride Manufacturing increase its customer base. The company estimated a 25% increase in the sales volume of fabricated parts if the lead time for the fabrication cell was reduced to 1 week.

Finally, Joseph wanted to use lessons learnt from improvements in the fabrication cell in other parts of the company. A low-volume, high-mix environment made lead time reduction an ideal approach for American Pride Manufacturing to improve its processes and subsequently increase its customer base.

With an eye on the external challenges faced by the company and the need to make changes in its business, Joseph decided to build a working team consisting of his few key managers to investigate improvement opportunities and then implement the changes. The shop had grown organically over the years and many of the existing processes needed a second look for a more efficient shop. Joseph had also noticed significant potential in one of his maintenance managers, Richard, who was also a recent hire. Richard had an undergraduate degree in Mechanical Engineering from Angelo State University and had a few years of experience in lean implementation

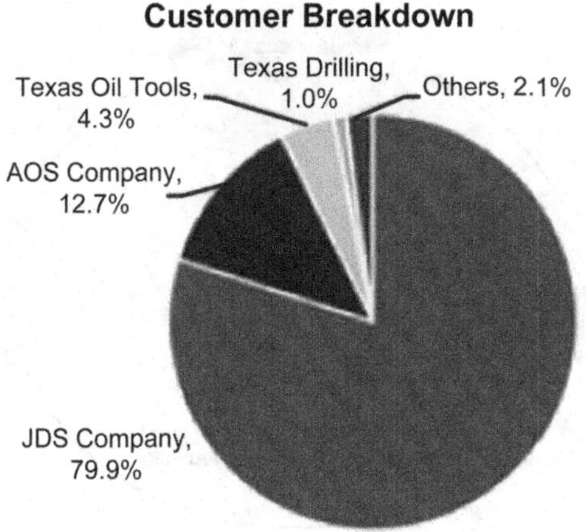

FIGURE 3.5 Customer breakdown by volume.

at his previous job. During their regular interactions, Richard had also highlighted a few issues that he had noticed at American Pride Manufacturing.

One of such insights that he had shared was related to service parts. American Pride Manufacturing had a total of 600 active part numbers, and it maintained a Finished Goods Inventory (FGI) of 363 parts. A significant number of these parts were service parts and were ordered in small quantities. These service parts were needed in much shorter lead time to minimize waiting for end customers. However, long lead time associated with these service parts necessitated maintaining an FGI for them. A shorter lead time would help the company reduce this inventory. Richard had analyzed finished goods inventory data and observed that a significant number of these were service parts. Joseph knew very well what had caused this buildup of service parts in the inventory. The existing Oil and Gas industry involved launch of new or improved tools and models in a 3 to 4-year period. Some of the parts from the old equipment model would become service parts which were ordered in smaller quantities and on an infrequent basis. The change in design of JDS parts and company's policy of running large batches to minimize the setups had resulted in a large number of service parts in the FGI. Figure 3.6 describes the inventory spiral associated with the FGI at American Pride Manufacturing.

A new part number is ordered in large quantity and is produced in the stamping area. However, this also means that the old design becomes a service part which stays in the FGI storage area for a long time. The service parts being ordered in smaller quantities could then be produced in the fabrication area. However, due to long lead time associated with the fabrication process, the service parts were still being produced in the stamping area. To save on setups, the number of service parts produced was higher than the actual demand and the excess units were stored in the FGI.

On a separate discussion with his purchasing manager, Lydia, Joseph also observed that long lead time associated with the raw material procurement was one of the major contributors of the long lead time. The sheet steel was ordered in batches of 10,000 lbs. with an average lead time of 3 weeks. The raw material for both the

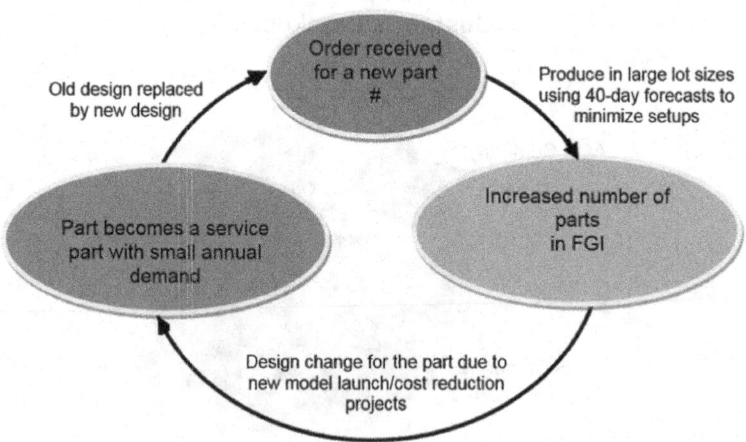

FIGURE 3.6 Inventory spiral.

stamping and fabrication processes was procured from six different suppliers in form of sheet steel. Lydia had also shared a preliminary analysis of order quantities across the six suppliers using the raw material purchase data for past 1 year. The data showed that approximately 90% of the sheet steel was procured from two distributors namely Steel Distributors and Bryce Steel located in Houston, TX. Together these two distributors supplied 51 different grades, thickness, and sizes of raw material. The raw material then stayed at the storage area for at least 1 week until the previous lot was used up.

Joseph noted that ordering raw material in larger batch sizes tied up company's working capital in inventory. With a shorter raw material procurement time, the company could order smaller batches more frequently, thereby reducing its working capital exposure. Furthermore, it would continue to allow the company to respond quickly to customer needs, but with much less cash tied to inventory on-hand.

After a few rounds of informal discussions with different managers, Joseph scheduled a formal kick-off meeting with maintenance manager Richard, manufacturing supervisor James, and purchasing manager Lydia. James was much older than Joseph and had always been Joseph's trusted advisor ever since he formally took over responsibilities of running the business from his father. James did not have a college degree, but his 25 years of experience with the company, extensive product knowledge, and a welcoming personality had made him one of the most valuable resources in the company. From his very early days, Richard had also developed a good rapport with him. Although Richard was much younger in age, James valued his expertise in ERP system and a general understanding of shop floor dynamics. Lydia had recently finished her ninth year with the company. She started as a buyer for high-volume raw material sheet steel parts and was later promoted to the role of purchasing manager. Lydia had the reputation of being a very dedicated employee and had known Joseph through family even before she started at the company. Joseph considered Lydia to be a very motivated person. A few years back he had encouraged Lydia to get her college degree through an evening class program at a nearby community college and then sponsored expenses related to her undergraduate degree in business management.

4 A New Approach to Fabrication

American Pride Manufacturing Turns the Page

The project team believed that identification of a clear project scope and focus was important to put a clear boundary around the project in such a way that it didn't try to achieve too much by keeping the scope broad, but at the same time made a sizeable impact on the business. To achieve the desired responsiveness for fabricated parts, using a cellular approach for making fabricated parts made most sense. The fabrication cell thus created would represent a small fraction of total production volume. The team believed that a quick success from implementation of lead time reduction in the fabrication area would not only help them get more confidence for future improvements but also get support from other employees in applying lessons learned from first project to the stamping area. Thus, the project's focus was on fabricated parts, and it included any part that used equipment in the fabrication cell. The project scope included raw material supply and shop floor operations including fabrication and welding. As an initial step, four machines were shortlisted for moving into the designated fabrication cell. However, equipment in the proposed cell were still being used for making parts for the stamping area. Furthermore, James pointed out that the painting operation took 2 days on an average, which is standard. As a result, painting was not included in the project scope.

To calculate product cost, many companies perform time study to estimate time taken per component and then multiply that time with the shop rate to get product cost. American Pride Manufacturing had also done a similar exercise many years back and the team still had access to that information. However, they did not have a system to track the total time it was taking a product from beginning to end. Shop floor data available from ERP systems are very useful in measuring and analyzing different aspects of production activities. However, most big brand ERP systems cost hundreds of thousands to millions of dollars to implement, making it cost-prohibitive for smaller companies like American Pride Manufacturing. The team believed that in order to set the project goal, they needed to know how long it was actually taking a fabricated part from start to end.

Richard suggested that one potential approach to address this limitation of not having system data could be to track operations using a traveler sheet for a smaller duration of time. After some initial discussion on what the team would like to measure, Richard led the exercise in the fabrication area to get an estimate of current lead time from start of production to delivery of the product. To implement this, Richard formulated a standard traveler sheet that could capture important job information. A sample traveler sheet is shown in Appendix A. Before initiating this exercise, Richard explained the objectives and details of the fields in the traveler sheet to all the concerned personnel in both office and shop floor. This was done to remove any misconception among the operators that the exercise was being used to measure their performance. The log helped to keep track of the sheets on the shop floor and ensure that no sheets were lost. An early check of the first lot of traveler sheets received from the shop floor revealed that a few of the fields were not being interpreted correctly by a small number of operators. This problem was corrected by explaining the details to the concerned employees.

Using data from the traveler sheets Richard created a lead time map for the team. From the lead time map, the total lead time for fabricated parts was calculated to be close to 36 days. It included 3 weeks of raw material supply and 1 week of queue time for raw materials and 8 calendar days of fabrication. Details of the breakdown of the lead time can be seen in Figure 4.1.

The current state lead time map showed that more than 73% of operations lead time was constituted by non-value adding activities and a 50% reduction target

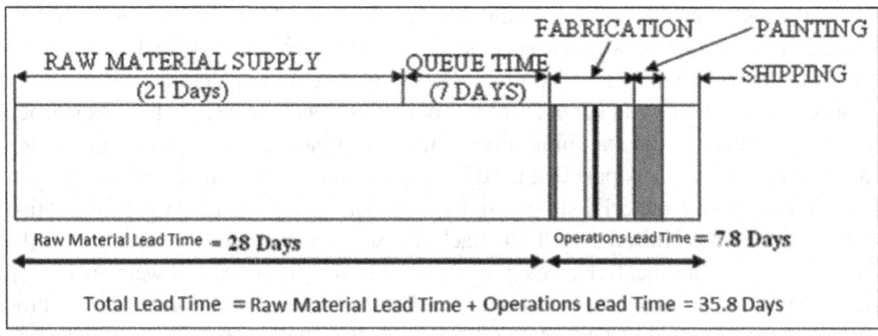

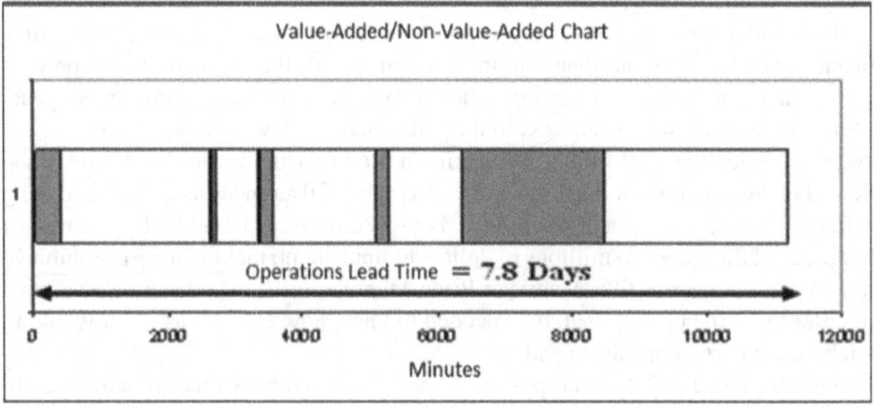

FIGURE 4.1 Current state lead time map.

TABLE 4.1
Goal for Lead Time

Focus Area	Lead Time	Current Lead Time (days)	Target Lead Time (days)	Percentage Reduction (%)
Raw material supply	Raw material	28	14	50
Fabrication	Operations	7.8	3.9	50

seemed feasible. Furthermore, through the analysis of current state raw material consumption rate and identification of suppliers with shorter lead time, it appeared possible to reduce raw material lead time by at least 50%.

Thus, the goal of this project was set to provide recommendations to reduce the overall lead time for the parts produced in fabrication cell by at least 50% from an average of 35.8 days to 17.9 days. Table 4.1 indicates the individualized target reduction for specified areas.

The scope of the project included activities from the time an order is placed for the raw materials to the time finished goods are shipped to the customer.

The team had broken down the responsibilities of collecting data and analyzing it with the intent of sharing insights with Joseph, as well as rest of the team members. There were several factors making this a challenging project. These included a somewhat interconnected manufacturing operation with multiple resources being used for products both in and out of project scope, as well as challenges in making physical changes to create a cell due to commitment to delivering on time to customers. Lastly, it was also important to win support from the employees by showing the impact of this project.

The team had gathered and analyzed different types of data during the project for insights and to develop a plan for change. Key data collected included sales history, manufacturing routers, lot-sizing policy, equipment reliability data, raw material purchase consumption rate, and supplier info. All the work started with meetings with relevant company personnel and data collection, followed by observation, root cause analysis, and concluded with specific recommendations for each problem.

Richard had used a process modeling software during his undergraduate studies many years back and volunteered to build one for this project. The team loved this idea because they could evaluate the impact of different recommendations before implementing it. Modeling would also help get the lead times and resource requirements, including labor and equipment, and utilization values for different possible alternatives. The overall project methodology that the team used is summarized in Figure 4.2.

To reduce or eliminate the Finished Goods Inventory (FGI) for service parts, it was necessary to reduce the lead time associated with the fabrication process so that they could be produced in the exact quantity and in a short time. Richard expanded on his previous analysis in this area by collecting weekly snapshots of FGI for a period of 11 weeks to observe the FGI for regular low-volume fabricated products. Figure 4.3 shows a comparison of the average FGI for 34 fabricated parts and their corresponding monthly demand in the same period.

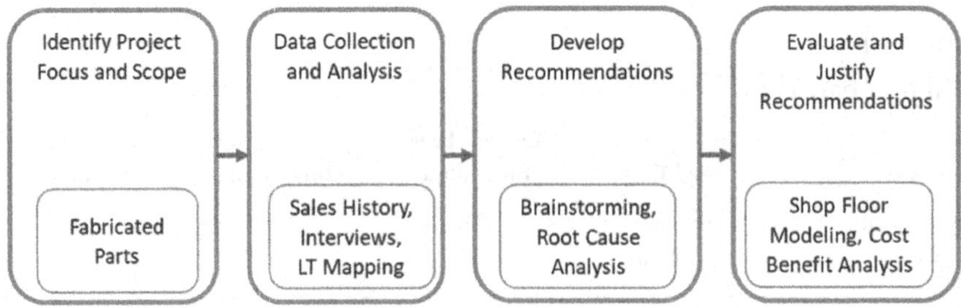

FIGURE 4.2 Project methodology.

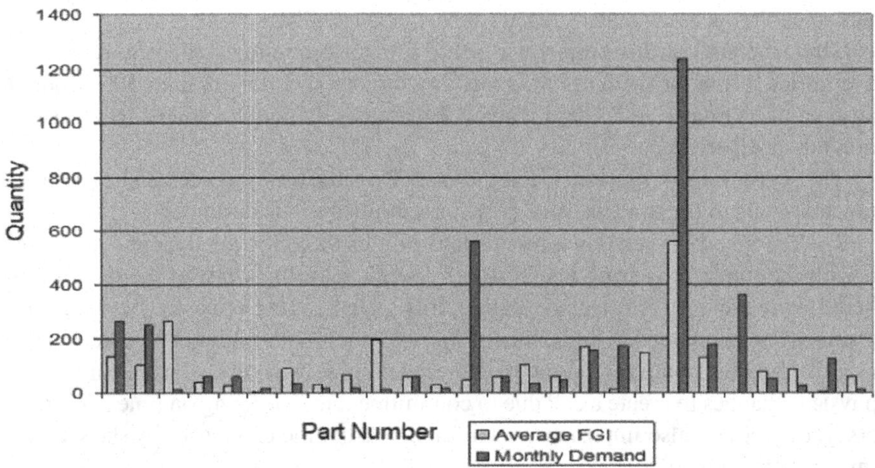

FIGURE 4.3 Average FGI versus monthly demand.

While presenting his findings with the rest of the team, Richard pointed out that the FGI is not directly related to the average monthly demand and is higher than the monthly demand for more than half of the products. Furthermore, the weekly snapshots of the FGI were also used to observe the number of weeks for which parts stay in the warehouse before an order is placed for them.

Furthermore, the graph in Figure 4.4 shows that on an average a part stayed for 7 weeks in the FGI before it is ordered by the customer. This confirmed that the FGI quantities were not accurately linked to the demand and were mostly driven by the mentality of running large batches to save on setups.

Instead of calculating more accurate storage quantities for the products in finished goods inventory, reduction in time associated with the fabrication process seemed to be a much better approach. The shop floor model developed by Richard had useful insights on lead time, Work-in-Process (WIP), and machine and operator utilizations. The output helped to achieve strategic analysis while maintaining a high-level view. To build the model, data such as demand history, Bills of Material (BOM), routings, machines, and number of operators were obtained from the company's ERP system. Richard realized that analysis would be much easier when the total number of the respective products modeled is less than 20. Since the actual number was close

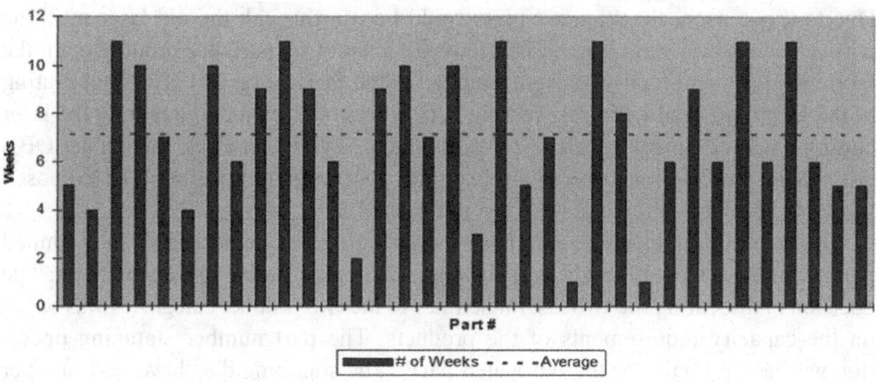

FIGURE 4.4 FGI holding period.

to 35, the parts were aggregated in 14 product families based on similarities in routings and operation times. To calculate the setup times and run times for the aggregated parts, a demand-weighted formula was used. Since the focus of this project was on fabricated parts, the stamped parts run through the laser and CNC punch machines were modeled as a set of dummy products to block capacity on existing machines for these parts. The dummy products were created by aggregating product in four different groups based on their routings.

The operators were modeled in four groups namely the laser operator, press brake operator, stamping press operator, and welding operator. The welding and press brake operator's availability for the fabricated parts was only 50% and Richard incorporated this in the model. The remaining time of these operators was used up for the stamping operations in the north plant. The lot size of all the fabricated products was taken to be equal to their average monthly demand. This was consistent with the current practice of producing parts for 4 weeks of demand. The details of labor and equipment modeling parameters are shown in Appendix B. The lead times weighted with the number of setups per year in this model was obtained to be slightly greater than 8 days, which was consistent with lead time map developed using travel sheets in the fabrication area.

The lead time chart from the modeling software explained the main reasons for the long lead time. It showed that the total time for any product mainly consisted of equipment setup, wait for labor, and wait for rest of the lot while the actual run time was negligibly small. This suggested that the reduction in lead time could be achieved through a reduction in lot size and by increasing the labor capacity. This was also confirmed through the labor utilization output, which showed high utilization levels of welding and press brake operators.

Armed with an understanding of root causes of long lead time, the next step was to evaluate potential of creating a fabrication cell. For this objective, James analyzed the manufacturing routers to identify the possibility of changes in routing of the products considered in the fabrication cell. The objective was to identify a self-contained product family that could be produced in the fabrication cell. Also, the stamped parts used sheet steel of less thickness as compared to the fabrication cell.

Due to this reason, the stamped parts could be run through the old laser machine with a lower power rating. James started with a list of 60 parts for production in the fabrication cell and finally 42 were included in the fabrication cell. The final routing of the identified product family required six pieces of equipment including the laser cutting machine, turret punch, 135T press brake, 35T Open Back Inclinable (OBI) press, MIG welding machine, and rocker arm spot welding machine. The proposed layout for the fabrication cell is shown in Figure 4.5.

The laser cutting machine with lower power rating was dedicated for the stamped parts while the one with higher capacity was dedicated for the fabricated parts. The decision of allocating the two laser machines to the two product categories was based on the capacity requirements of the products. The part number stamping operation was needed only for the fabricated parts. The stamping dies have part number impression machined on their inner surface and therefore the stamped parts do not need part number stamping as a separate operation. This made it easier to dedicate the 35T OBI press to the fabrication cell. The relocation of the 35T press in the fabrication area would also make transfer batching possible. In addition, the collocation of equipment would facilitate material handling and help reduce the existing policy of running large batches to minimize material handling across the two plants.

The development of this fabrication cell required a lot of collaboration between James and Richard while Joseph also participated in those meetings to share his insights. Once James identified a potential list of parts along with their new routings, Richard developed another model to evaluate the capacity requirement for the proposed fabrication cell. The new model included the products with new routings as well as the stamped parts which needed at least one operation outside the fabrication cell. As mentioned earlier, 42 out of 60 parts did not need any equipment outside the fabrication cell. The remaining 18 products needed one of the presses in the stamping area. These 18 products were allocated on the 2600 W laser while the remaining products were allocated on the 4000 W laser. A dedicated time slot for

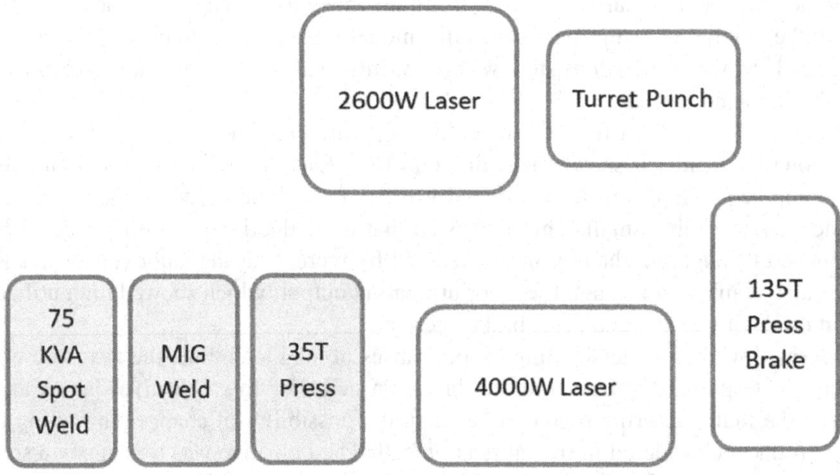

FIGURE 4.5 Proposed layout of the fabrication cell.

fabricated parts was proposed for the presses that were needed for the 18 products in the fabrication cell. All the equipment in the fabrication cell were modeled for one shift, but the laser machines were modeled to run for two shifts. A total of three operators were modeled in the fabrication cell with one operator assigned to the two laser cutting machines, CNC punch, press brake, and the 35T OBI press in first shift. The second operator was assigned only for welding operation on the spot and MIG welding machines. The third operator was assigned to run the two laser machines in the second shift. The details of labor and equipment modeling parameters in the proposed case are shown in Appendix C. The output in the new model showed a major reduction in lead time, but a major portion of the total flow time still consisted of wait for the rest of the lot. The next step, therefore, was to perform multiple simulations in the model to identify lot sizes that resulted in minimum flow time. Although, reduction in lot size for different products caused the number of setups to increase, this was more than offset by a reduction in waiting time for the lot. This simulation was eye-opening for James also, who had always relied on the traditional approach of getting more parts per setup to increase efficiency.

In the final model, lead time weighted with the number of setups per year was equal to 3.7 days which was a 52.5% reduction. The utilization level for the 2600 W laser cutting machine was observed to be significantly higher than that of 4000 W laser because of high-volume stamped parts run through them.

Based on questions from Joseph and James, Richard also performed additional analysis using the modeling software to identify the cell performance in case of an increase in production volume of the fabricated parts. The analysis showed that for an increase of 25% in demand of all the part numbers, the fabrication cell would still be capable of producing parts with a weighted lead time of 3.25 days. However, this would need rerouting of 20% of the workload from the 2600 W laser to the 4000 W laser, as well as the addition of a welding operator. This analysis was useful in convincing Joseph and James that they will not have to undergo another major change in the production process in case of a demand increase.

On the raw material procurement front, purchasing manager Lydia suggested starting with two key suppliers, since they represented a high percentage of total raw material procurement. Lydia also performed additional analysis related to the distribution of order quantities across the 51 types of raw material from these two suppliers. Figure 4.6 shows the percentage distribution by weight for different types of sheet steel.

Lydia pointed out that out of 51 varieties of sheet steel, 14 accounted for nearly 76% of the total raw material procured from these two suppliers. To narrow down the list of raw materials, it made sense to focus on these 14 types of raw material. Lydia also analyzed the distribution of the ordering frequency across these 51 types of raw material and the percentage of annual demand constituted by each subgroup as shown in Figure 4.7. The chart showed that nearly 76% of the total raw material was ordered less than once a month and therefore indicated that there was a good opportunity for reducing the on-hand inventory.

During the early raw material-focused discussion session, the team observed that when the time spent by raw material was demand weighted, it showed that the raw material spends nearly 40 days on the shop floor before the next order was placed.

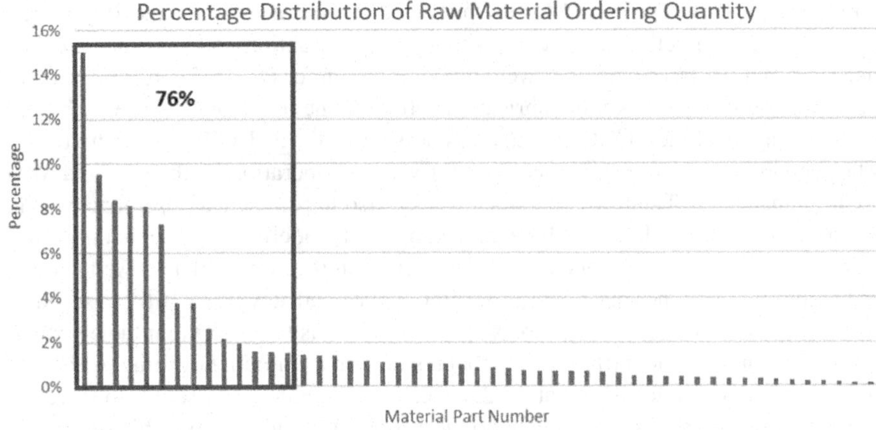

FIGURE 4.6 Percentage distribution of ordering quantity by weight.

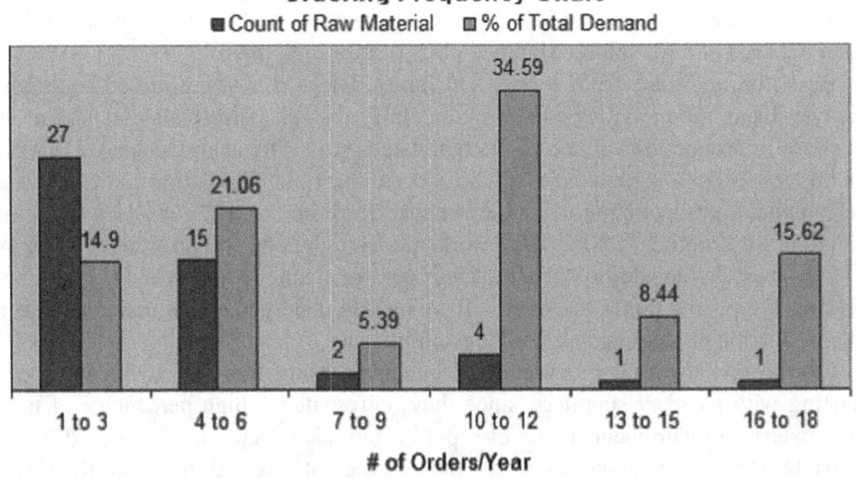

FIGURE 4.7 Frequency distribution chart of order.

This showed that there was an opportunity to reduce on-hand inventory. Lydia and Joseph followed up with these two suppliers and learned that the identified grades of steel were high selling grades and were stocked by the distributor in large quantities. It was also realized that the identified grades of steel could be procured in smaller quantities with a lead time of 3 days if American Pride Manufacturing could give an estimate of its raw material requirement for a month-long period.

The team evaluated two possible alternatives to reduce raw material inventory and associated lead time. The first approach was to use historical data on raw material procurement and implement a reorder point. This approach involves use of averages for calculating the order quantity and reorder point, but it is more applicable in case of a stable demand. The Reorder Point (ROP) for raw materials is calculated using the following formula:

$$\text{ROP} = \text{Average demand in replenishment period} + \text{Safety stock}$$
$$= d \times LT + Z \times S_d \times \sqrt{LT}$$

where d is the demand in the replenishment period and LT is supplier lead time. Z is a constant that corresponds to the desired service level. A statistical Z-table is used to get the Z-value corresponding to a specific service level and is equal to 2.63 for a 99% service level. Also, S_d is the standard deviation of weekly demand.

The second alternative was to share current 40-day forecast with the supplier and order raw material on a weekly or biweekly basis. The 40-day rolling forecast would help the distributor estimate the raw material requirement thereby eliminating the chances of stock-out resulting from high variation in raw material usage.

Among the two alternatives discussed earlier, Lydia believed that the second approach of sharing the forecast with the distributor was better due to two key reasons. Although the ROP approach is more data-driven, it comes with the additional challenge of updating the calculations on a frequent basis. Furthermore, this recalculation will also be needed if new fabricated items are added to the mix. For the approach of sharing 40-day forecast, further discussion was carried out with raw material distributors, Steel Distributors and Bryce Steel. The discussion resulted in consensus on a new raw material procurement strategy where American Pride Manufacturing would share its forecast for the 8-week period with the two suppliers while the raw material will be procured on a weekly basis.

Based on all analysis performed and alternatives evaluated, James, Lydia, and Richard presented a detailed plan to the company owner Joseph. Their final recommendations were focused on the following three key areas:

1. Creation of a fabrication cell and change from make-to-stock to make-to-order approach.
2. Implementation of a new raw material procurement strategy.
3. Reduction of FGI for fabricated parts.

Further details of these recommendations are discussed below:

1. **Creation of a Fabrication Cell**:
   i. **Move Equipment**: As shown in Figure 3.3, the fabricated parts had to move across the two buildings several times during the fabrication process. To reduce material handling, James recommended relocation of the 35T OBI, MIG welding machine, and the spot welding machine into the fabrication area. This would result in a significant reduction in material handling and facilitate the change in lot-sizing policy. In addition, James and Richard used the insights from the simulation model to recommend dedication of the 2600 W laser cutting machine to parts from the stamping area and dedicate the 4000 W laser to the fabrication cell. The fabrication cell thus created would have six equipment

including the laser cutting machine, turret punch, 135T press brake, 35T OBI press, MIG welding machine, and rocker arm spot welding machine.

ii. **Change Part Routing**: To create an isolated family of the fabricated parts, the routings of a few parts were changed. The parts that were earlier being run through the 90T and 70T press brakes in the stamping area were assigned to the 135T press brake in the fabrication cell. Also, all the parts produced in the fabrication cell used either the 4000 W laser or the CNC punch as the first station while the sheet metal for the stamped parts that was initially cut on the 4000 W laser was later assigned to run through the 2600 W laser.

iii. **Cross-Train Operators**: The insights from simulation model were also helpful in identifying labor needs and their allocation on different machines. James had picked three operators for the fabrication cell with one operator on first shift assigned to the two laser cutting machines, CNC punch, press brake, and the 35T OBI press. The second operator was assigned on first shift only for the welding operation on the spot and metal inert gas (MIG) welding machines. The third operator was assigned to the two laser machines in the second shift. The two laser operators in the fabrication cell were recommended for cross-training to run the press brake and the 35T OBI press. While this is unusual in many big companies, since the operators are primarily dedicated to one machine, James believed that the small size of the company combined with the favorable attitude of the three operators would allow him to implement this change. The welding operator was going to get trained to operate both spot as well as MIG welding machine.

iv. **Implement Make-to-Order Strategy**: The team asked Joseph to move from make-to-stock to make-to-order strategy for the fabricated parts. The make-to-stock strategy had resulted from long lead time associated with the fabrication cell. Also, the lot sizes were based on the monthly forecast of the fabricated parts. A shorter lead time and the ability to run smaller lot size would enable the company to implement a make-to-order policy. In addition, implementation of transfer batching would also be needed in the fabrication cell.

2. **Implementation of the New Raw Material Procurement Strategy**: Based on her discussions with raw material suppliers, Lydia recommended to share the current 40-day forecast with the two suppliers namely Steel Distributors and Bryce Steel and order the raw material on a weekly basis. The 40-day rolling forecast would help the supplier estimate raw material requirement thereby eliminating the chances of stock-out resulting from high variation in raw material usage. Figure 4.8 shows the schedule recommended for ordering the raw material.

The new ordering process would involve American Pride Manufacturing providing the 8-week forecast to its distributors followed by regular updates of RMI requirements after every 4 weeks. The distributor would hold

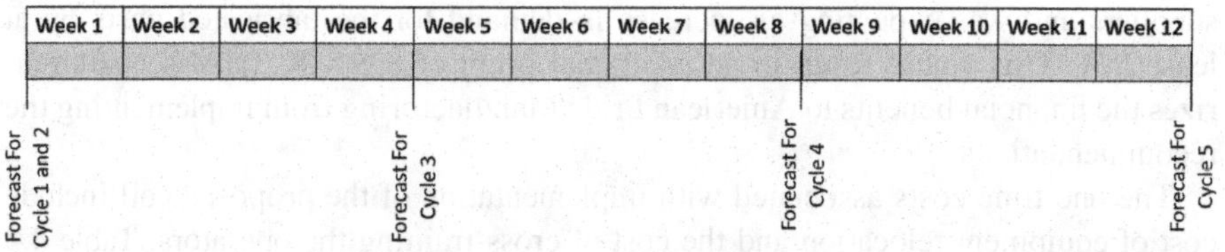

Week 1	Week 2	Week 3	Week 4	Week 5	Week 6	Week 7	Week 8	Week 9	Week 10	Week 11	Week 12

Forecast For Cycle 1 and 2
Forecast For Cycle 3
Forecast For Cycle 4
Forecast For Cycle 5

FIGURE 4.8 Proposed raw material ordering strategy.

inventory for a cycle of 4-week period while American Pride Manufacturing would procure its raw material once a week. Since the raw material was shared by stamping and fabrication areas, production of a week's demand in the stamping area was suggested so that excess production would not result in the stock-out of raw material. At the same time, a new lead time reduction project was initiated in the stamping area to control the impact of frequent setups on the delivery performance of stamped parts. The procurement policy for the rest of the raw material was left unchanged.

3. **Elimination of FGI**:
   A major portion of FGI at American Pride Manufacturing consisted of fabricated parts. The reduced lead time of the fabricated parts would ensure that orders for low-volume parts could be delivered quickly to customers. The team, therefore, recommended a gradual elimination of the FGI for such parts. However, to safeguard against any unexpected events, the team agreed with Joseph that a safety stock of 1 week's demand be maintained only for high-volume products in the fabrication cell. In addition, the team suggested to shift production of low-volume service parts from the stamping area to the fabrication area. The team strongly believed that all these recommendations would help American Pride Manufacturing break out of the inventory spiral discussed in the previous section.

Joseph was actively involved in this project at different stages and even attended most of the team meetings. By the time, final recommendations were being finalized, he also wanted to make sure that all these recommendations made financial sense as well. So, he asked his accounting manager to work with James, Richard, and Lydia to develop a cost-benefit analysis. The team used calculations from the simulation model and raw material usage analysis to conclude that American Pride Manufacturing would realize a one-time benefit from reduction in WIP, RMI, and FGI. The reduction in WIP was estimated to result in a one-time saving of \$1397. This was estimated by comparing the simulation model output for WIP in the base case and the recommended case. In addition, the implementation of weekly procurement strategy for 14 types of raw material would cause a significant reduction in the amount of raw material on-hand. This was estimated to result in a one-time saving of \$13,869. The reduction in FGI was estimated to result in a one-time saving of \$18,385.

A reduced lead time for the fabricated parts was also expected to help American Pride Manufacturing increase its sales by expanding its customer base. Joseph had listed American Pride Manufacturing in two online marketplaces for buyers and suppliers and was expecting an increase in demand for the fabricated parts by at least 25%. This would result in an additional profit of $39,848. Table 4.2 summarizes the financial benefits to American Pride Manufacturing from implementing the recommendations.

The one-time costs associated with implementation of the proposed cell include cost of equipment relocation and the cost of cross-training the operators. Table 4.3 summarizes one-time costs.

Being a first project of this scale and type at the company, Joseph wanted these recommendations to be implemented effectively. So he asked the team to also present a detailed implementation plan. The team's proposal was based on an eight-step implementation plan as shown in Table 4.4, and it covered the rollout of recommendations in three areas including creation of the fabrication cell, raw material ordering strategy, and the FGI management.

Before implementing any recommendation, Joseph and the team believed it will be best to get the relevant people educated about the benefits of cellular layout as well as other changes being implemented. This is because the success of any project relies heavily on change in the mindset of the workforce performing their job responsibilities. This was especially important for the operators who would be impacted when improvements on the shop floor would be implemented. Changes like cross-training and task reallocation require better teamwork and mutual understanding.

It also made sense to start the physical changes with space allocation for part number stamping press and welding machines in the fabrication area. Since American Pride Manufacturing had a unionized workforce, the decision to create the

TABLE 4.2
Summary of the Financial Benefits

One-Time Benefits	
Raw Material Inventory Reduction	$13,869
WIP reduction	$1397
Finished goods inventory reduction	$18,385
Total one-time benefits	$33,651
Annual Benefits	
Profit from increase in sales (25%)	$39,848

TABLE 4.3
Summary of Financial Benefits

One-Time Cost	
Equipment installation	$3000
Training	$2000
Total One-Time Cost	$5000

TABLE 4.4
Implementation Timeline

Steps	Tasks/Activities	Duration	Weeks 1	2	3	4	5	6	7	8
1	Inform employees on the benefits of changes	1 week								
2	Floor space allocation for the stamping machine	1 week								
3	Installation of stamping machine	2 weeks								
4	Floor space allocation for welding machines	1 week								
5	Installation of welding machines	2 weeks								
6	Implement the new raw material ordering policy	4 weeks								
7	Production trial for the new cell	3 weeks								
8	Start cross-training of personnel in fabrication cell	6 weeks								

fabrication cell was preceded by discussions with the union representatives regarding the need and strategy for project implementation. Visible support from the top management and involvement of the union representatives in the discussions helped reduce initial resistance to change. In addition, James worked with one of his staff to develop standard operating procedures in the fabrication area to ensure that different cross-trained operators followed the same standards. This was quickly followed up with the relocation of two welding machines and stamping machine and by cross-training of the operators in the fabrication cell.

Simultaneously, Lydia started the new raw material ordering policy with the two suppliers. This was accompanied by a change in the production policy for the stamped parts. As mentioned earlier, this change in production policy was necessary, as the raw materials were shared by both stamping and fabrication areas. Since the available raw material was tied to the weekly demand, any excessive production in the stamping area would result in stock-out of the raw material. Therefore, Joseph initiated another improvement project in the stamping area to reduce the lead time for high-volume stamped parts. This was necessary as the new production policy in the stamping area would have affected the on-time delivery performance of stamped parts.

Once the desired increase in capacity level was achieved through the operator relocation and recommended changes in layout, the next step was to run the first production trial in the fabrication cell. The reduction in lot size is usually not favored by the operators due to the traditional mindset of maximizing production on each setup. Therefore, James suggested implementing batch size reduction in three equal steps in a time span of 3 weeks. Since this had a direct impact on the lead time, it was always important to get success in the first trial to get the buy-in from the operators. Joseph also led a companywide communication of success in initial lead time reduction efforts to create awareness about the success of this initiative toward lead time reduction. Finally, Joseph also worked with accounting lead to develop and implement a lead time measurement system for American Pride Manufacturing to track improvements in each of the identified areas.

The transition toward a different approach to produce parts at American Pride Manufacturing has many unique features that made it successful. This starts from direct involvement of the company owner Joseph Schultz. Next, the team he built had the right mix of skillset, experience, and a sense of ownership toward their respective areas of work. By combining their individual strengths with a vision for change and by keeping the scope narrow enough to improve his chances of success, Joseph was able to start a new chapter toward making American Pride Manufacturing better prepared to compete with low-cost competitors.

5 Gaining Market Share through Shorter Lead Time

The Case of a Colorado Shop

CCB Products Corporation (CCB) was an enclosure solutions and custom contract manufacturing business located in a 500,000 square foot facility in Boulder, Colorado, for the past 70 years. The company designed, manufactured, and assembled a wide range of standard and configurable cabinets, enclosures, frames, racks, and mounting products. These products were used in data communications, telecommunications, audio-visual broadcasting, and security applications. These products were sold directly to OEM, integrators, and end-users. CCB had a union facility with a highly skilled workforce of 200 employees; 68% of which had over 25 years of experience, with 50% having more than 30 years of experience. In the years preceding the project, the company had faced the challenge of several ownership changes, with the most recent one resulting from an acquisition by a private equity firm. Alex Stolz was the newly minted Operations Director and was charged with turning around operations to capture a bigger market share.

The products manufactured at CCB could be classified into two broad categories, standard and custom. The standard products include preconfigured cabinets, consoles, frames, and rack mounts with options for different sizes. Standard products were mostly made to stock, and Finished Goods Inventory (FGI) was replenished after a product was shipped to the customer. Custom products on the other hand were made to order according to the customer specifications. These specifications are provided by the customer in two different ways: either in the form of drawings of the required product or specifications of the design changes required in an existing standard product.

The manufacturing process flow for both custom and standard products is summarized in Figure 5.1. The process started with the receipt of a customer order. For a custom product, these orders were first sent to design engineering for detailed product and process requirements while orders for standard products are directly entered into CCB's ERP system. MRP was run twice a day, once in the morning and then again in the afternoon to ensure quick feedback on newly entered orders and start the planning process. The "time-phased open orders" report generated by the MRP system was then utilized by the manufacturing manager and operators to fabricate parts for orders. Both standard and custom products mainly used sheet steel and aluminum or steel plates as raw material. The fabrication was completed in two distinct areas within the facility namely, standard and custom manufacturing. Both lines

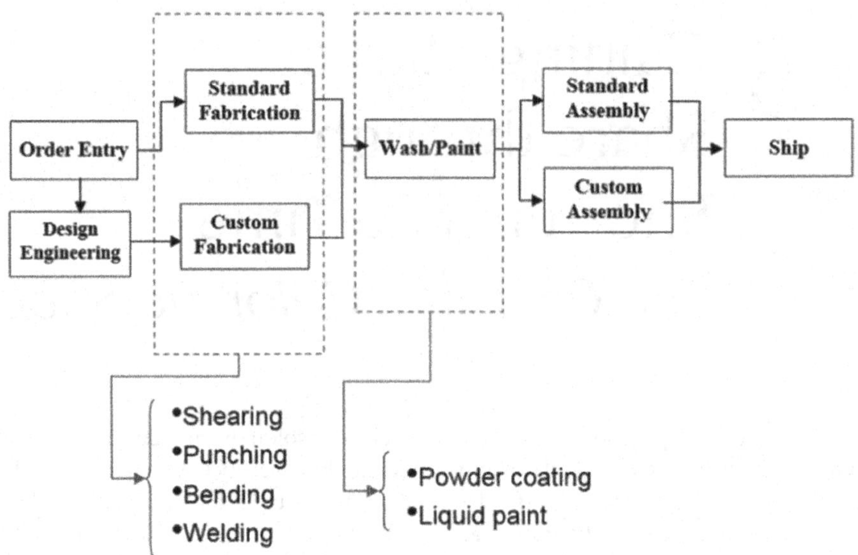

FIGURE 5.1 High-level process map.

had a functional layout and used NC punch presses, brake presses, and manual and automatic welding stations for shearing, punching, bending, and welding of parts, respectively. However, all products shared order entry, paint, and shipping resources with other product lines. The manufactured parts were then picked from their inventory locations by material handling and moved to paint. This was done according to the "time-phased open orders" report and paint sequencing rules for a sales order number, product type, and color. The parts are then loaded on to the conveyor, which went through the wash, paint, and assembly lines. The wash and paint line consisted of manual and automated stations while the assembly operators use hand tools to complete specified assembly operations. Assemblies were then packaged along with kits for a sales order and shipped to the customer.

CCB had several reasons for reducing lead time in its facility. In competitive business markets, companies no longer competed only on quality but on responsiveness and product variety. In addition to this, there was increased competition from low-cost countries manufacturing similar products. Alex saw this as the main challenge; he knew he had the talent, but needed to start rethinking processes. Prior to 2001, CCB was the market leader in their industry but has since struggled to be competitive and had fallen between fifth to seventh out of 12 companies in their market. Products from low-cost countries had also started flooding the market in recent years; this was threatening the competitive position of the current players. From Alex's perspective, if things did not change quickly, the plant would have to be shuttered in the long run. The company estimated its external lead time (from order entry to product shipping) at 7 days for standard products and 42 days for custom contract products. In addition, CCB faced long supplier lead times ranging from 5 to 7 weeks for supply of most of the raw materials needed for manufacturing its products. However, customers in CCB's industry expected manufacturers to be much more responsive to their needs.

CCB's competitors had competed on lead time by stocking inventory in warehouses close to customers and have effectively reduced their external lead time to one day. CCB did not have the resources to match their competitors' strategy, so adopting a short lead time strategy would have given them the opportunity to become competitive and responsive, while reducing inventory and other costs. In addition to struggling to compete on the lead time, the organization was also not very competitive on product variety for their standard products. They had 10,000 active part numbers for components and wanted to more than double the number of components to over 23,000 to meet customer needs quickly.

Alex started thinking about a product line where he could launch an initiative to improve responsiveness and use that as a model to drive improvement in other product lines. With that aim in mind, he started analyzing the business. Figure 5.2 depicts the business breakdown by sales for CCB, illustrating that the custom products equated for 65% of the total revenue, whereas the standard products equated to only 35%.

The custom products had a 42-day lead time, which seemed significant compared to the standard products at 7 days, but 42 days was acceptable by industry standards, and more importantly CCB's customers.

The total market size for the standard product was approximately $500 million of which CCB had a share of approximately $10–$15 million. As a result, Alex believed this product line had the greatest potential for growth. Furthermore, a major percentage of the custom product orders received at CCB involved modifications in an existing standard product. Alex believed that a reduced lead time for standard products would enable it to provide more options for customization in standard products category, thereby increasing its percentage in total sales volume.

The standard products were once again broken down into two main categories, the Prime and Pride products, and the Printed Circuit Board Enclosures (PCBE) and Cellular Tower Cable Racks (CTCR). Figure 5.3 illustrates the breakdown by sales and volume, respectively, for these products within the standard product line.

The Prime and Pride products represent majority of the sales in the standard products category, but the PCBE and CTCR products make up majority of the production volume. In addition, CCB personnel explained that the organization currently holds

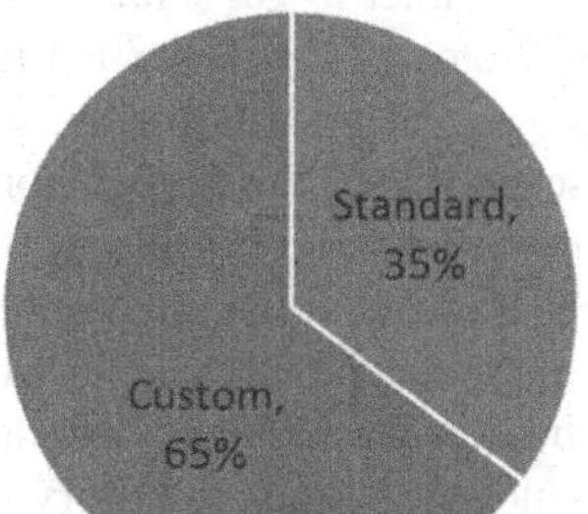

FIGURE 5.2 CCB business breakdown.

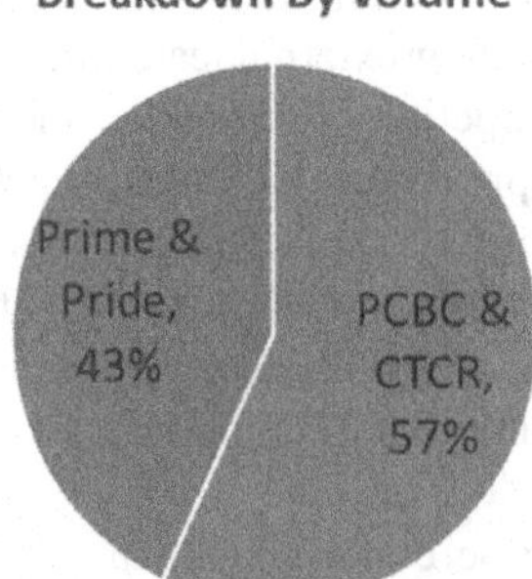

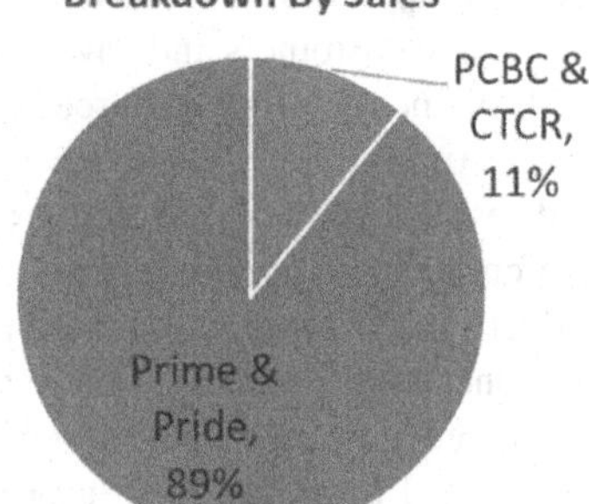

FIGURE 5.3 Breakdown of product types by sales and volume.

less than 5% of the total market share for these product types, but there has been a significant increase in demand in this category. The company's sales team expected that with a reduced lead time, CCB could expect a significant increase in sales revenue. From a manufacturing standpoint, this category of product used dedicated equipment; so, Alex decided to form a project team to make improvements in this product line and replicate the changes as they successfully implemented the changes in the PCBE and CTCR product line.

The project team started mapping out processes in all areas in the factory and collecting ERP system data in parallel. Processing orders at CCB started with the receipt of orders. Over 30% of all the incoming orders contained incomplete or inaccurate information. Customer service personnel had to call the customer or the salesperson back to gather the necessary information to complete the order. This process involved long waiting periods, with manual follow-up on paperwork accumulating on every individual's desk.

The customer service group consisted of three employees. Timely order entry was critical to signal manufacturing to start processing and ensure that the order left per date promised in the sales order acknowledgement. Average order entry time was around 10 minutes, but it had a huge amount of variation even on standard orders because required fields were left empty by customers or sales personnel entering the purchase order, who used their own order entry formats.

CCB also experienced high variability in their demand. There were no contracts in place with regular customers either to get some insight into demand patterns or quickly react to their needs. Demand variability led to inconsistent workload for customer service representatives and manufacturing, with cycles of low workload followed by overtime work several times during the month. Figure 5.4 captures representative demand variability at the time of the project.

Raw material procurement was one of the major areas for opportunity to reduce lead time at CCB because most material had a 5–7-week lead time through their current suppliers. Not only did this cause CCB to forecast their sales out 5–7 weeks, but the company needed to hold safety stock inventory to meet forecast uncertainty. However, despite high inventory levels, the long lead time still caused stock-out situations.

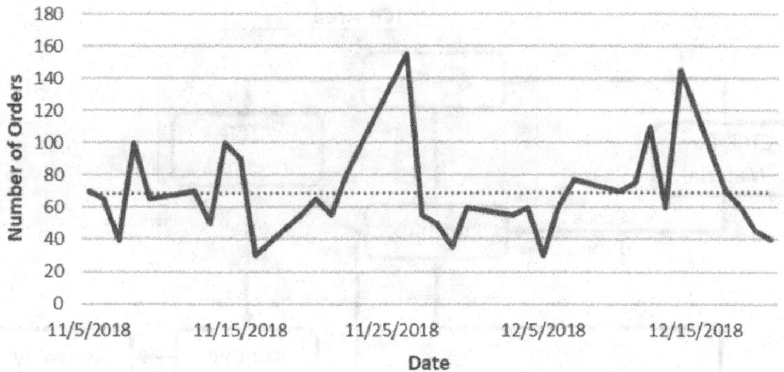

FIGURE 5.4 Daily orders for standard products.

PCBE product was a two-post open rack system which was used for IT, datacenter, networking, computing, and storage applications. Figure 5.5 shows the representative bill of material for the PCBE product. It consisted of three components: top and base face, door, and side walls.

The Cellular Tower Cable Rack (CTCR) product was a cable management product, and it prevented tangles and damage to cables, thereby maintaining quality data transmission. A CTCR consisted of two components: side tube and cross tube as shown in Figure 5.6. Each CTCR required 2 side tubes and 6to –11 cross tubes depending on length.

An overview of the current process flow in manufacturing is shown in Figure 5.7. As mentioned earlier, the CTCR and PCBE manufacturing areas were not collocated.

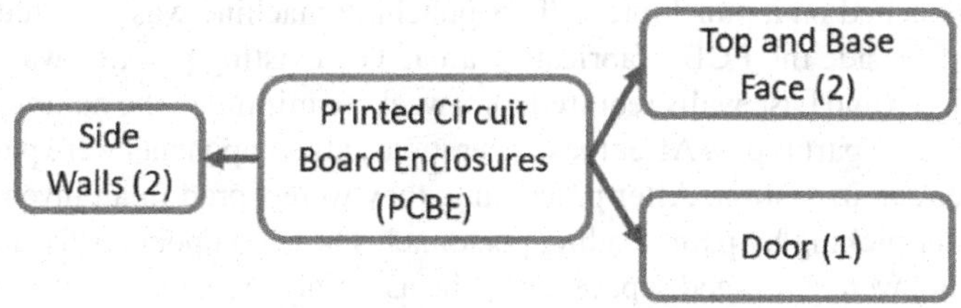

FIGURE 5.5 Bill of material for PCBE.

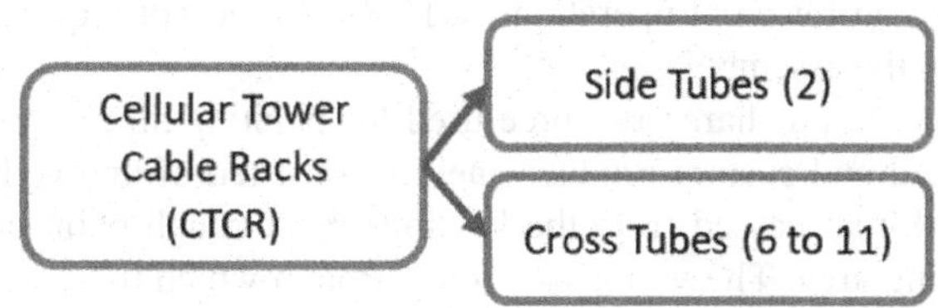

FIGURE 5.6 Bill of material for CTCR.

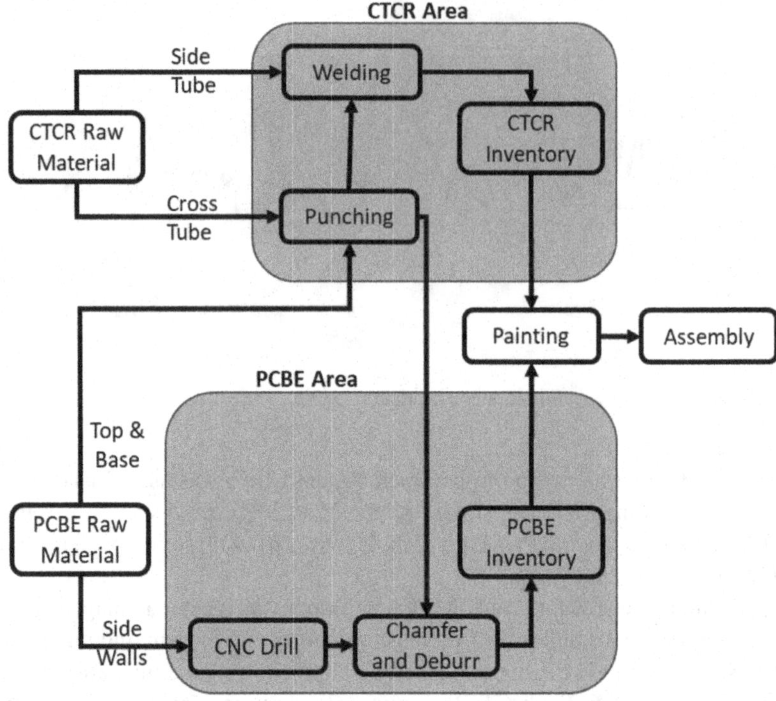

FIGURE 5.7 Overview of existing process flow.

The raw materials for PCBE came cut to size from the supplier. Holes in the side walls were drilled using a CNC drill while holes of same size on the top and base walls are punched on a punch press. This punching machine was a shared resource and located outside the PCBE fabrication area. The existing practice was to run an entire lot of top and base walls required for a week, to minimize the number of setups needed for these part types. After these operations, all components were passed on to the chamfer/deburr station. After processing, they were stored in an inventory location for order picking by paint loading personnel. These components did not require assembly and were packaged separately in the assembly area.

The raw materials for CTCR also came cut to size from the supplier. The cross tubes were punched on both sides on the CTCR punch and then welded to the side tubes on a welding fixture. Welded CTCRs were stored at an inventory location and subsequently picked for paint operations. CTCR also did not require assembly and was packaged in the assembly area.

The paint line was a shared resource used for painting all the products coming from the standard and custom product lines. It consisted of an overhead conveyor, 2500 ft. in length that passed from the loading area through paint onto the assembly and packaging area. The various stations through which the conveyor passed is shown in Figure 5.8.

The operation in the paint area started with part loading. The part loading personnel picked up parts from the fabrication area based on shipment dates and stored

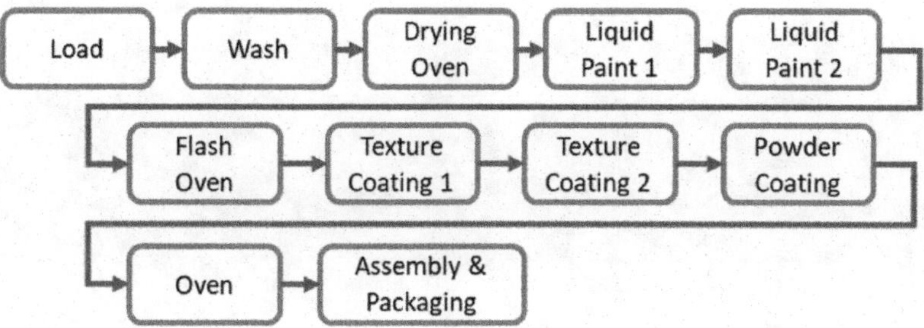

FIGURE 5.8 Overview of operations flow in paint and assembly.

them in the part loading area. Parts were then loaded onto the conveyor hooks based on the colors being run on that day. This process minimized the number of setups for color change; CCB considered a time window of 5 days and all the parts of a specific color that require shipping within that 5-day period were painted in a single lot. Once loaded onto the conveyors, a part passed through all the stations until they were offloaded at the assembly station. Cycle time for the conveyor was at 4.8 h.

The four highlighted stations shown in Figure 5.8 used 320 ft. of conveyor length and were used only for wet paint operations. However, wet painting comprised only 5% of the total volume of paint operations.

Alex sat down with his team and reviewed the initial findings. Preliminary discussions were centered around creation of a roadmap for the project with the selected product line. All personnel agreed they needed to have a quick win to start moving the facility on a journey where responsiveness to customer needs was the culture of the organization.

6 CCB Takes a New Approach to Become More Responsive

Preliminary work on the project started with data collection which included historical sales data, interviews with both office personnel as well as line operators, randomized sampling, along with individual observations. Through this preliminary analysis and discussion, a current state value stream map was established. This map offers valuable insights into the lead time components through the system. Table 6.1 shows the current state lead times for the PCBE and CTCR products.

Alex and his team saw that the operations lead time was substantially longer than CCB's quoted lead time. One of the main reasons for this was CCB's current lead time was achieved using work-in-process and finished goods running in the system.

After baselining the current state, the project team started outlining ideas for improvement using tools like brainstorming sessions, nominal group technique, and cause-effect diagram. Alex led these discussions to gain insights from some of the key functional areas.

The results of all discussions were transferred to a cause-and-effect diagram as shown in Appendix D. This further helped narrow down the focus of the project into five major areas and identified possible alternatives for achieving the target lead time as shown in Table 6.2.

The baseline lead times were tabulated based on the value stream map and targets were developed based on theoretical calculations. Table 6.3 captures all data for this exercise.

The project team targeted an average of 85% lead time reduction in the Cellular Tower Cable Rack (CTCR) and Printed Circuit Board Enclosure (PCBE) areas. These reductions would be achieved by improving the five areas listed in Table 6.2. Alex instructed the project team to start gathering detailed time study data for lead time analysis in the manufacturing lines. The ERP system provided data related to historical sales, manufacturing routers, purchased material pricing, and sources.

TABLE 6.1
Current State Lead Time

CTCR	PCBE
Total lead time = 129 days	Total lead time = 162 days
Quoted lead time = 29 days	Quoted lead time = 88 days

TABLE 6.2
Improvement Areas and Possible Alternatives for Target Lead Time

Area of Improvement	Alternatives for Achieving Target Lead Time
Customer order flow (order receipt and order entry)	Use Robotic Process Automation (RPA) Simplify order entry
Purchased/Raw material flow	Dual suppliers Share forecast data with partner suppliers In-house sizing
PCBE fabrication flow (setup, fabrication, WIP)	Cell creation Cross-train operators
CTCR fabrication flow (setup, fabrication, WIP)	Cell creation Cross-train operators
Pre-paint, WIP, and Scheduling flow	Scheduling procedures Paint alternatives

TABLE 6.3
Current versus Target Lead Time for Identified Improvement Areas

Areas of Improvement	Current LT (days)	Target LT (days)
Customer order flow (order receipt and order entry)	1.5	0.5
CTCR purchased/raw material flow	99	21
PCBE purchased/raw material flow	87	21
CTCR fabrication flow (WIP, fabrication, weld, paint)	27	2
PCBE fabrication flow (WIP, fabrication, paint)	72	2

A data collection form for time study data and inventory count in the fabrication area was distributed to the shop leads. The inventory count obtained at each station and the average production rate obtained from the study of sales records helped develop the current state lead time map using Little's Law, which states

$$\text{WIP} = \text{Lead} - \text{time} \times \text{Production Rate}$$

Process flow study and data collection were used to understand the process flow and analyze the causes for long lead time. Different sets of data collected from the ERP system included processing time at each station, lot-sizing policy, number of equipment and operators, size and location of equipment, inventory level at each station, and production volume. This dataset was collected to model the shop floor operations and understand improvement opportunities in the existing process.

The project team studied the raw materials used for manufacturing the PCBE and CTCR products. The raw material CCB used was T6005-T6 aluminum for the PCBE frames and cold-rolled steel for the CTCRs. Upon comparing material type for competitors' products with CCBs, it appeared that the competition was using the same steel, but T6061-T6 aluminum. T6061-T6 aluminum was the industry standard and was readably available through a distributor.

Different distributors were contacted to get quotes on the materials. While talking to a quoting representative from one of the distributors, it was evident that the cross-sections of some of the raw materials procured were not the same as industry standards. Since a distributor provides stock length raw material from a warehouse and does not usually carry specialized raw materials, it would not be possible to get raw materials as per exact dimensions needed for the PCBE side walls. The flanges on the channel were 1.250″; however, distributors were only able to provide channels with flanges with a length of 1.410″. The current dimensions of the CTCR cross tubes were 1.000″×0.5000″×0.060″, distributors were only able to provide rectangular tubing that was 1.500″×0.500″×0.065″.

For standard materials, most distributors quoted a 2–3 day lead time compared to CCB's current supplier lead time of 5–7 weeks. The closest distributor quoted a lead time of 2 days if the material needed was in the distributor's Denver, Colorado, warehouse. If the material needed to be shipped from the Dallas, Texas, warehouse, the lead time will be 3–4 days.

Table 6.4 shows the different material configurations that distributors were able to quote and pricing. The PCBE base face was not included in the quote because its cross-section was not standard; thus, it was not stocked by the distributors. Other options were explored for the PCBE base face.

It is important to note that the price per piece for the PCBE top face and CTCR cross tubes were cheaper through the distributor. Due to the significantly higher price per piece for the PCBE side walls, it was determined that other options may need to be explored for this part.

The lead time for the PCBE base face and the PCBE side walls from an alternative supplier was quoted at 3 weeks compared to 5–7 weeks with their existing supplier. The prices of each of the pieces can be found below in Table 6.5.

The prices for a few of the components in Tables 6.4 and 6.5 were more expensive than their current prices. However, a shorter supplier lead time would result in a

TABLE 6.4
Information Acquired from Distributors for Raw Materials

Material	Length (ft)	Qty.	Quoted Price per Piece ($)	Current Price per Piece ($)
PCBE side walls	50	50	25.51	11.63
PCBE top face	25	25	1.21	1.43
CTCR side tube	50	50	3.96	4.56
CTCR cross tube	50	50	0.60	0.36

TABLE 6.5
Information Acquired from Alternative Supplier for Raw Materials

Material	Quoted Price per Piece	Current Price per Piece
PCBE base face	$9.26	$7.94
PCBE side walls	$12.11	$11.63

significant inventory reduction of raw materials and a lower the cost tied up with this inventory, thereby compensating for the increase in raw material prices. The project team also proposed creating a pool of partner suppliers with whom CCB could share raw material requirements data and put a risk-sharing agreement in place. The partner supplier would receive a time-phased planned order report every week from CCB and procure all material that falls within the agreed delivery lead time window and put policies in place for ordering material outside the delivery window based on material commonality, conversion ease, and other risks based on material type. This agreement would allow the partner supplier to react quickly to CCB's needs and lower the risk of stocking material in excess of CCB's needs. It was proposed that a quarterly review be conducted with the supplier to keep inventory in line with future and current requirements. The meeting would also facilitate exchange of information regarding increasing future business activity and allow the supplier to take some steps internally before actual requirements showed up on the reports from CCB.

In order to reduce the lead time for PCBE and CTCR fabrication processes, Alex proposed the following approach to create a single cell consisting of PCBE and CTCR fabrication area. The overview of proposed process flow is shown in Figure 6.1.

The proposed fabrication cell would consist of collocated PCBE and CTCR fabrication area. The raw materials for top/bottom angle for PCBE and the side and cross tubes for CTCR in the proposed cell would be procured from a distributor and cut to size within the plant facility using a CNC band saw. One new band saw would have to be purchased by CCB. The existing band saw available at CCB would require some

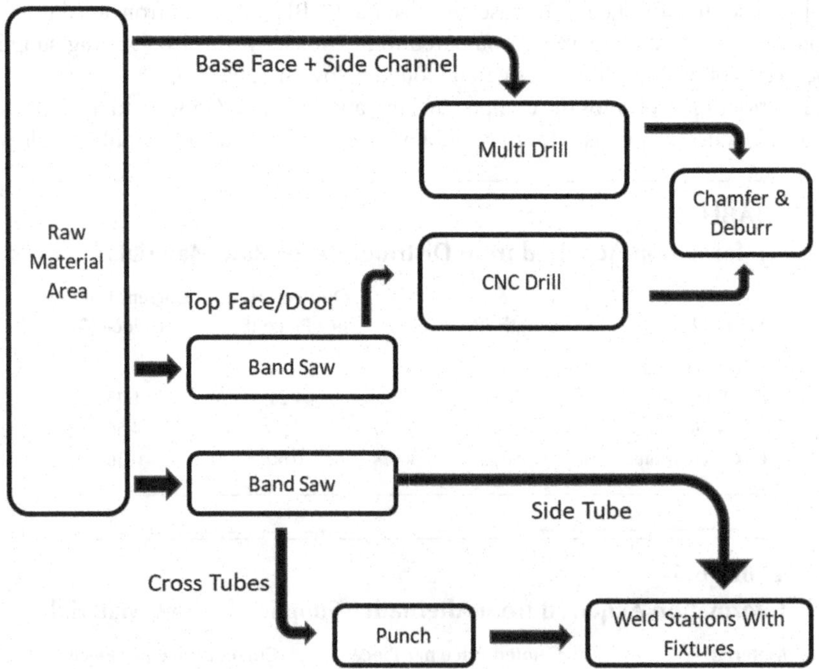

FIGURE 6.1 Overview of proposed process flow in cell.

modification for cutting multiple pieces of raw material at a time. This change could be accomplished by the internal maintenance department. Using an existing press for sizing and punching operations was also considered, but it was not adopted due to the following reasons:

- A large number of setups were required for different operations (shearing and punching of different raw materials), thereby reducing cell flexibility.
- Relocation of punch into the proposed cell was not possible due to constraint arising from low roof height in the fabrication area.

The bottom face and side channels (PCBE component) would be procured cut to size from the supplier. Kanban cards would be set up between the supplier and CCB to allow ordering of material without an actual purchase order being issued. Alex assisted the project team with the development of the Kanban cards and the necessary calculations for the number of cards that needed to flow between the supplier and CCB. The PCBE area would use a multidrill in place of punch for making holes on top and bottom face.

The manufacturing engineering team started to analyze the capabilities of the proposed cell, data collected from the fabrication area, and sales records were used to build a baseline simulated model of the proposed cell with four basic products, namely PCBE and three varieties of CTCR. Paint operation was also integrated into the model. The model was built assuming a timeframe of 250 working days per year and one shift per day.

Once the process simulation was developed, and the base model replicated the performance of the current system, several alternatives were evaluated for equipment requirements, labor allocation, and lot-sizing policies to reduce lead time. The simulation showed that it was possible to achieve a lead time of 1 day in the CTCR and PCBE areas using the labor allocation shown in Table 6.6. However, one extra labor would be required in the proposed cell for operating the CNC band saws.

The simulation output also showed requirements for equipment for the proposed cell; details are shown in Table 6.7. CNC OP and CTCR OP represent labor allocated in PCBE and CTCR fabrication areas, respectively. The proposed cell would need two CNC band saws for sizing all the raw materials.

Several alternative scenarios were evaluated in the simulation and the utilization level of the equipment indicated that the proposed cell would have sufficient capacity to meet future increases in demand. The lead time estimates obtained from the simulation illustrated that it was possible to achieve a lead time of 1 day for both PCBE and CTCR fabrication with the cellular layout. After developing the model for the

TABLE 6.6
Labor Allocation Summary

Labor Group	Count	Comment
PCBE Area (CNC OP)	2	1 Extra labor needed
CTCR Area (CTCR OP)	1	Same as current

TABLE 6.7
Equipment Requirements

Equipment	Number	Labor Allotted	Comments
Band saw	2	CNC OP	1 New equipment
Chamfer	1	CNC OP	Existing equipment
CNC drill	1	CNC OP	Existing equipment
Multidrill	1	CNC OP	Existing equipment
Punch	1	CTCR OP	Existing equipment
Radius fixture	1	CTCR OP	Existing equipment
Weld fixture	1	CTCR OP	Existing equipment

current year, additional analysis was performed to estimate the operator and equipment requirement with increasing demand for a period of 5 years. Additional optimization was performed to understand equipment and labor requirements for medium and lower levels of demand over the same period. In every case, modeling the cell showed that it was possible to achieve a 1-day lead time for both PCBE and CTCR fabrication. The simulation results looked very encouraging to Alex; the prospect of shorter lead time in this product line would help keep the organization aligned for future projects for other product lines.

Pre-paint WIP observed in the production process indicated that additional analysis was needed to reduce the waiting time for paint operation.

In order to reduce the lead time for paint, the existing operating details were studied to estimate the capacity of the paint line. This was done to evaluate the possibility of reducing the scheduling time window from 5 days to 1 day. Alex reached out to Scott, the supervisor in the paint area, to evaluate the paint line capacity.

Scott proposed estimating the current capacity of the paint line using the following steps.

1. Orders received for the past year collected from order entry.
2. Orders sorted based on color.
3. Total number of yokes needed per item.
4. Number of yokes left free for loading and color change.

The other inputs needed for capacity estimation are shown in Table 6.8.

Alex got with the project team and performed the following capacity calculation.

Capacity Calculations

1. Time between two successive loadings = Distance between yokes/ conveyor speed = 0.24 min
2. Number of yokes available in an 8 h shift = Total time available for loading/ Part loading rate = {8(h/day) × 60(min/h) – 5 (setups/day) × 6(min/setup)} / 0.24 (min) = 1875
3. Number of yokes required (estimated from orders received in Dec 2018) Max: 1582; Min: 1149

4. Utilization of paint line = (Number of yokes required/day)/ (Number of yokes available/day)
   Max. utilization: 84.4%;
   Min. utilization: 61.3%

From the utilization values obtained above, it was clear that the paint line had sufficient capacity to paint all the parts for orders received in a single day.

Furthermore, various suppliers were contacted to understand possible improvement opportunities in the paint area. One of the alternatives to speed up processing in the paint area was through the use of a gas catalytic infrared oven instead of the gas-fired oven currently used for the paint curing operation after powder coating. Installation of the new infrared oven would reduce the cycle time by 1 h. An additional advantage of having shorter curing time was the possibility to increase the conveyor speed from 8.33 fpm to 12 fpm. There were alternatives discussed about segregating the powder coating from the wet painting line as well. The estimated cycle time reduction with the various possible options in the paint area is shown in Table 6.9.

TABLE 6.8
Inputs for Capacity Estimation of Paint line

Setup time	6.00 min
Conveyor speed	8.33 fpm
Distance between yokes	2.00 ft
Number of setups each day	5
Hours of operation	8.00 h
Total length of the conveyor	2400 ft

TABLE 6.9
Cycle Time Calculations for Paint Line

	Conveyor Length (ft)	Conveyor Speed (fpm)	Current Cycle Time = Conveyor Length/Conveyor Speed	Percentage Reduction in Cycle Time
Existing Paint line	2400	8.33	2400 / (8.33×60) = 4.80 h	
With infrared oven	(2400 – 475) = 1925	8.33	1925 / (8.33×60) = 3.85 h	19.8%
With infrared oven and increased conveyor speed	(2400 – 475) = 1925	12	1925 / (12×60) = 2.67	44.4%
With infrared oven, increased conveyor speed, and segregation of wet paint line	(2400 – 475 – 140) = 1785	12	1785 / (12×60) = 2.48 hrs	48.4%

TABLE 6.10
Capacity Calculations for Paint Line

	Time between Two Successive Loadings = Distance between Yokes/ Conveyor Speed	Number of Yokes Available in an 8 h Shift = Total Time Available for Loading / Part Loading Rate	Percentage Increase in Capacity
Existing paint line	0.24 min	{8(h/day)×60(min/h) – 5 (setups/ day)×6(min/setup)} / 0.24 (min) = 1875	
With increased conveyor speed	0.17 min	{8(h/day)×60(min/h) – 5 (setups/ day)×6(min/setup)} / 0.17 (min) = 2700	44%

Furthermore, with an increased conveyor speed, the capacity of the paint line could be increased by 44% as shown in Table 6.10.

After conducting a thorough analysis and modeling the future state using manufacturing simulation, the project recommendations focused on the following:

1. Increase order entry responsiveness to eliminate the backlog situation.
2. Improve supplier/raw material flexibility to reduce supplier lead times
3. Create a PCBE/CTCR fabrication cell and convert from make-to-stock to make-to-order.
4. Streamline paint operations and implement a 1-day schedule.

The recommendations for order entry department focused around increasing the order entry's responsiveness to customer order and eliminate the backlog situation. The first recommendation was to create and deploy a standard order reference sheet. Customer service representatives mentioned that nearly one-third of all the orders they receive required customer callbacks due to lack of and/or conflicting information on the order. This demanded a significant amount of time trying to communicate with the customer and re-process existing orders. The customers received a products catalog from CCB with part numbers and accessories, but no standard order form. These customers then emailed in their own customized purchase order respective to their company's format. The project team worked with order entry personnel and created a standard order reference sheet. This sheet formed the basis of creating a web page for customers, to enter order requirements in a standard format that could be quickly processed by CCB. It was recommended that the order reference sheet be added in CCB's catalog, along with the web address, and have sales representatives discuss it directly with the customer.

The project team also investigated the use of Robotic Process Automation (RPA). RPA is a technology that is governed by logic and structured sequential inputs, aimed at making business processes automatic. Using this tool, a company can interpret applications for completing transactions, manipulate data, trigger responses, and communicate with other digital systems. The basic process would involve personnel performing their day-to-day tasks, but the system would run in the background learning all inputs over some duration depending on task complexity. After the training period for the system is over, the system would perform the task and there would be a control tower to quality check the output of the task. As mistakes get corrected through the control tower, the system's accuracy to complete the task would improve. The control tower tasks would be maintained by the best performing employee in the department who would monitor and improve the performance of the system. The ability to use RPA would allow employees to switch their focus to more thoughtful and meaningful work while also eliminating data-entry errors that can slow processing times, compliance, and the overall customer experience.

The project team proposed the use of RPA for the regular orders and to slowly scale it up to manage more complex orders that would involve interaction with sales personnel, possibly using a chatbot. Using this approach, 50% of the regular order entry time would be freed up allowing personnel to focus on more complex orders.

The project team put together a cost analysis tool for management to evaluate the financial impact of the RPA implementation shown in Table 6.11. The analysis included the following input parameters:

Tasks the RPA bot would automate: 3
Number of employees currently performing task(s): 3
Percentage of the employee's daily time spent on these tasks: 50%
Employee average salary: $50,000/year

The calculated payback for this investment will be 1.75 years.

Another recommendation for increasing order entry responsiveness was to improve CCB's website and align it with CCB's business objectives. First, the company's website was intended to be used as a tool to view and familiarize potential customers with the organization's products, but, in its current state, it did not support this. Second, it was suggested that the website also include a product configurator tool. This would allow customers to research the products and accessories more in-depth. Two of CCB's main competitors created their configurator tool and were utilizing this tool for their customers. This led to exact products and part numbers associated with them, which greatly reduced the number of inaccurate or contradicting information on customer orders.

The recommendations for raw material procurement focused on improving supplier/raw material flexibility to reduce supplier lead times. The first recommendation was to switch from CCB's current suppliers to either new suppliers or contract with standard material distributors with a focus on lead time reduction.

Regarding the raw material supply chain, Alex suggested developing partner relationships with multiple distributors and few specialized raw material suppliers. With contracts in place, suppliers with access to planned order information from CCB's ERP system could flex volumes and respond to CCB's needs quickly. The project team modeled demand and supply using advanced machine learning techniques and shared the outputs with Alex. Using these advanced techniques, CCB could forecast demand and drive planning bills within their ERP system which would drive raw

TABLE 6.11
RPA Cost Justification Calculations

	Year 0	Year 1	Year 2	Year 3	Year 4	Year 5	Total
Current employee costs		$1,50,000	$1,50,000	$1,50,000	$1,50,000	$1,50,000	$7,50,000
Employee hours/ year that will be automated with RPA		3000	3000	3000	3000	3000	
Employee cost/ year that will be automated with RPA		$75,000	$75,000	$75,000	$75,000	$75,000	$3,75,000
Cost to implement RPA bot	$1,05,000	$15,000	$15,000	$15,000	$15,000	$15,000	$1,80,000
Net cash flows	($1,05,000)	$60,000	$60,000	$60,000	$60,000	$60,000	
Cumulative cash flow	($1,05,000)	($45,000)	$15,000	$75,000	$1,35,000	$1,95,000	
Tasks the RPA bot would automate:	3						
Number of employees currently perform the task(s):	3						
Percentage of the employee's daily time spent on these tasks:	50%						
Employee average salary:	$50,000						
Payback (Years)	1.75						

material demand. This mechanism would give CCB additional flexibility to have orders in process before the receipt of an actual order. The forecasts would be consumed when an actual order was placed by the customer.

Several long-term recommendations were also provided to further improve supplier/raw material flexibility. First, it was recommended that CCB continue to contact distributors for quotes and explore further opportunities for lead time and price reduction. The second recommendation was to investigate switching to industry standards for all the raw materials. This included the PCBE base face, PCBE side walls, and CTCR cross tubes. If industry standards were used for these pieces, all material could be procured from a distributor and the lead times would be reduced even further. In doing so, CCB would have control over sizing material to length. This would provide CCB with the ability to increase their product variety and even provide custom lengths at no added cost. Finally, the project team recommended changing the raw material for PCBE to an industry standard T6061-T6 aluminum. This material being an industry standard would enable CCB to source all of its raw material for PCBE through a distributor. Alex assigned personnel from the manufacturing engineering team to understand testing requirements that would allow for this switchover and evaluate design impact.

Equipment requirement and labor allocation obtained from manufacturing simulation modeling were used to develop the new layout of the fabrication area, shown in Figure 6.2. The new layout required collocation of the equipment of both CTCR and PCBE cells. A short-term solution was to allocate the two CNC band saws in the final relocation area and subsequently move other equipment.

The proposed cell would have two band saws operated by a single operator. The CNC drill and the multispindle drill were to be operated by a single operator in the PCBE area, while the punch and CTCR welding were to be handled by a single operator in the CTCR area. The arrows in proposed layout below indicate the material flow directions.

In the updated layout, the raw material for top face and door for PCBE would be in the form of standard-length aluminum plate purchased from the distributor while the side channel and base face would be procured cut to size from a supplier. The raw material for PCBE top face and door would be cut to the required length in the CNC

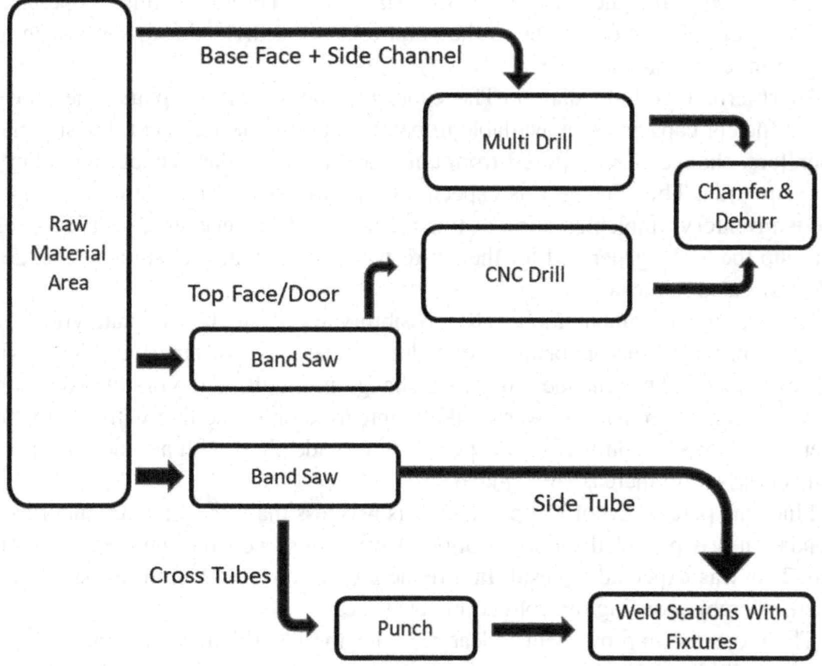

FIGURE 6.2 Proposed layout of CTCR/PCBE fabrication cell.

band saws and then transferred for further processing. Holes on the base face and side channel would be drilled on the multispindle drill station while the top face and door would be drilled in the CNC drill machine. After this, they would be moved on to the chamfer and deburr station. Machined parts would be kept in a rack for order picking by paint loading personnel.

The raw material for ladder rack consisted of steel tubes purchased from a distributor. Tubes would be sized on the band saw to form side tubes and cross tubes. The cross tubes would be punched in the subsequent operations and welded to the side tubes.

The next recommendation was to cross-train operators in the proposed cell. The operators from both PCBE and CTCR areas needed to be cross-trained to work on all equipment, during demand peaks of either CTCR or PCBE, operators from one area could be moved to another. Cross-training the operator would lead to job rotation, job satisfaction, team ownership, and coverage for absences.

In addition to cross-training, it was recommended to have a champion in the proposed cell for cross-training other operators during the low demand periods. The existing operator in the PCBE area was an ideal candidate for this position because of her experience in both welding and machining operations. Alex appointed the Manufacturing Manager to start negotiations with the union to start setting up the cell.

The project team recommended implementing standard operating procedures. The sequence of operations to be performed for producing a product needed to be standardized and implemented throughout the cell. Standard worksheets and inspection procedures needed to be displayed in the cell and operators trained to perform in accordance. The implementation of the recommended layout was expected to result in reduced lead time due to improved material flow and cross-training of operators.

The main recommendations in the paint area were divided into short-term and long-term recommendations.

Short-term recommendation: The capacity analysis of the paint line showed that sufficient capacity was available to paint parts that needed next day shipment. Therefore, change was proposed from a five day to single day scheduling policy in the paint area. This change was expected to eliminate the safety stock of finished goods inventory. Implementation of this recommendation needed direct incorporation into the MRP generated for the entire fabrication area, so a suitable change in MRP was also suggested.

Long-term recommendation: The possibility of using the gas catalytic infrared oven provided the opportunity of reducing the conveyor length and increasing conveyor speed. This change required a change in the drive motors of the existing conveyor system so that the oven could be integrated into the line without stopping operations. This option was also expected to provide CCB with additional capacity to meet the future increase in demand.

Since the percentage of wet painted parts was less than 5%, an additional recommendation was provided for segregation of wet paint line from the powder coating line. This was expected to result in a reduction of conveyor length to the extent of 120 ft., thereby reducing the conveyor cycle time.

The projected improvement in lead time for the identified improvement areas is shown in Table 6.12. The baseline lead time and target lead time from the project goal are also shown for planned versus actual comparison.

The project team developed an advanced machine learning model to capture the demand pattern and forecast future demand for the standard products. External influences like metal prices, cloud technology trends, data center setups, and IoT adoption were factored into the models. Several iterations were carried out by the team before selecting the appropriate model. The chosen model root mean squared error (RMSE) for the training data was low and running several test data samples showed favorable results. This model was adapted for forecasting demand and recording accuracy for future adjustment purposes. During the project period, the accuracy of the forecast was evaluated, and the error rate was in the range of ±8%. The customer order flow recommendations for using RPA and forecasting were expected to significantly shorten the lead time.

Tables 6.13 and 6.14 summarize the target and projected lead time that will be achieved through the implementation of the raw material recommendations.

TABLE 6.12
Projected Improvement

Area of Improvement	Baseline LT	Target LT	Projected LT
Customer order flow (order receipt and order entry)	1.5 days	0.5 day	1 day
CTCR purchased/raw material flow	99 days	21 days	7 days
PCBE purchased/raw material flow	87 days	21 days	35 days
CTCR fabrication flow (WIP, fabrication, weld, paint)	27 days	2 days	2 days
PCBE fabrication flow (WIP, fabrication, paint)	72 days	2 days	2 days

TABLE 6.13
Target Lead Time

CTCR	PCBE
Total LT = 25.5 days	Total LT = 25.5 days
80% reduction	84% reduction
Quoted LT = 4.5 days	Quoted LT = 4.5 days
85% reduction	94% reduction

TABLE 6.14
Projected Lead Time

CTCR	PCBE
Total LT = 12 days	Total LT = 40 days
91% reduction	75% reduction
Quoted LT = 5 days	Quoted LT = 5 days
84% reduction	93% reduction

The project team performed an analysis of the sales over a 5-year period. The year over year growth in sales was estimated to be 15% (see Appendix 1E-Sales Growth Estimate). A summary of results is shown in Table 6.15.

The one-time costs associated with the implementation of proposed cell included the cost of new equipment, equipment relocation cost, and die tooling costs. In-house sizing of the raw materials for PCBE and CTCR required purchase of one new band saw in current year at $60,000. Retooling the existing band saw, tooling costs with vendor, and equipment relocation would add an additional $16,860 in project costs. Operation of the band saw in the cell would need an additional operator as well. Labor costs were estimated based on the existing market labor rate.

The supplier selection for raw materials was based on comparing the supplier lead time reduction and percent cost difference. Colorado Extrusion Company, a new extrusion supplier, and another distributor were used for comparison purposes. The percent cost difference and lead times are summarized in Table 6.16.

For the PCBE top face, a distributor was selected because it had both lower costs and lead times. Colorado Extrusion Company was selected for the side walls and base face. The base was not available at the distributor, thus establishing a critical-path lead time of 3 weeks for PCBE. As a result of the previously established critical-path lead time and higher cost at the distributor for the side tubes, Colorado Extrusion Company was selected.

The cost of changing raw material suppliers, to reduce lead time and ensure availability to support surges in demand, was justified by comparing the piece price quotes from the current supplier to that of prices from Colorado Extrusion Company and the distributor.

The following four quantifiable benefits are detailed.

TABLE 6.15
Summary Cost versus Benefit Analysis

Payback Period (Pre-Tax)	2.3 Years
Net present value	$341,539
Internal rate of return	53%

TABLE 6.16
Percent Cost Increase and Lead Time Comparison for Raw Materials

Product Description		Colorado Extrusion Company		Distributor	
PCBE	Top	31.6%	3 weeks	−18.1%	2–3 days
	Side	3.9%	3 weeks	53.4%	2–3 days
	Base	14.3%	3 weeks	–	2–3 days
CTCR	Side	−43.0%	3 weeks	−15.2%	2–3 days
	Cross	51.4%	3 weeks	39.8%	2–3 days

One-Time Inventory Investment Savings: A significant benefit to implementing the project recommendations were the reduction in inventory. This included the following

a. Reduction of raw material on-hand.
b. Reduction of work-in-process (WIP) stock.
c. Reduction of finished goods stock.

The savings from inventory reduction were estimated by comparing the WIP levels obtained from model simulation and initial shop floor inventory count; it amounted to $45,788 (Appendix F: One-Time Inventory Investment Savings).

Annual Holding Cost Savings: Closely tied to the one-time inventory investment savings was the annual holding cost savings. $45,788 one-time inventory savings × (20% Holding cost – 10% WACC Factor = 10%) resulted in an annual holding cost savings of $4,579 per year.

Raw Material Supplier Salvage Savings: The savings attributed to implementing the recommended raw material supplier changes came specifically from the selected distributor. The PCBE top face and CTCR side and cross tubes would be cut to length, while the resulting scrap would be salvageable. Salvage material costs were evaluated at 10% of the actual purchased material cost (on a per lb. basis).

Sales Revenue Increase: The most significant benefit of implementing the project recommendations and reducing lead time was the potential sales revenue increase. CCB's sales team expected that with a reduced lead time, CCB could expect sales revenue to increase by at least 15% year over year for a period of 5 years.

Summary of the total cost-benefit is shown in Appendix G: Total Cost-Benefit Schedule. This analysis showed the payback period for the project to be 2.3 years.

Alex expected several challenges during implementation; he started summarizing his thoughts.

1. While considering a vendor for implementing RPA, it was emphasized that one size does not fit all organization or all its needs. Selection of the vendor for easy and rapid scalability to accommodate transaction volume spikes would be a big advantage for RPA. Any vendor with a high-end infrastructure will take less deployment time and less turnaround time for scalability purposes.
2. While implementing RPA, it will not be difficult to see that automation will have widespread implication for employees. Operations would be required to invest time in defining new roles and responsibilities for staff whose roles would be impacted as a result of this change. Management would also have to choose implementation functional areas, set expectations for the team managing the implementation, create operational realignment, and service governance.

3. After investing in RPA automation, the performance measures and accuracy standards will need to be redefined. Personnel involved in managing the control tower need to have standards that continuously improve the system's capability to perform more complex tasks over time.
4. Implementing the machine learning forecasting model will require monitoring and tuning as business conditions change. The planning team may need IT support to maintain a functional model.
5. Changing raw material required CCB to perform all the required tests on raw material. Without testing, it would be difficult to switch to a different material (T6061-T6) or to change the material cross-sections that were being used currently. Timelines need to be added to the project plan to accommodate material testing.
6. Investment in new equipment was justified, based on increased future sales from lead time reduction. CCB would need to leverage the organization's future ability to increase product variety and pursue lead time reduction goals, increase sales year over year to keep the project economics alive.
7. CCB had a functional layout and implementation of the recommendations needed manufacturing reorganization to allow the creation of a cell with collocated PCBE and CTCR fabrication areas. This allocated area was needed prior to the purchase of the band saw. In addition, the proposed layout required enough space to allow the transportation, storage, and sizing operation of standard-length plates on a band saw. The existing layout would hamper movement of the longer length raw material purchased from the distributor.
8. CCB had a union workforce; implementing a cell and cross-training the workforce required a major change in the mindset and division of labor. CCB would have to work with the union to create suitable job profiles for cell operations. Operators and management also required training to develop a lead time reduction–based mindset to better understand ownership issues related to creation and running the cell on a day-to-day basis. Establishing standard operating procedures starting from the new cell and creating a plan to implement it companywide were also needed. Top management needed to get personnel involved in the change initiative.
9. Implementing a one-day schedule needed incorporation into the MRP generated for the entire fabrication area, so a suitable change in MRP was needed. It was recommended to make these changes in steps to win employee support and allow time for issues to be uncovered and resolved.
10. The installation of infrared oven in future needed trials to determine the operating parameters which may interfere with the regular production. Furthermore, the paint personnel would have to be trained for operation and maintenance of the equipment. Speeding up the conveyor had the potential for reallocation or need for additional operators in future.
11. Segregation of wet paint area from the powder coating line needed a separate area for the wet paint booths, as well as necessary arrangements for material handling.

TABLE 6.17
Implementation Timeline

Steps	Tasks/ Activities	Duration	Month 1	Month 2	Month 3
1	Educate employees on benefits of project	2 weeks			
2	Identify and cross-train personnel for order entry	3 weeks			
3	Contact recommended and new suppliers	1 month			
4	Floor space allocation for cell creation	2 weeks			
5	Equipment purchase and commissioning	1 month			
6	Selection of cell team members	2 weeks			
7	Detailed cell planning	1 week			
8	Existing equipment relocation	3 weeks			
9	First production trial for new cell	1 week			
10	Planning of changes in MRP for paint line	1.5 months			
11	Implement required changes in MRP	1 month			

Alex realized that the implementation of any project relies heavily on the people involved and their readiness to change their mindset in performing the job responsibilities. In the case of CCB, this included using smaller batch sizes, efficient scheduling, and proper material handling among other aspects. This also required training and educating the workforce in the company on the benefits of shorter lead time to expedite the implementation process.

CCB needed to start with the purchase of band saw equipment, clear floor space, and implement the recommended cellular layout for PCBE and LR. Employees had to be selected for the cells and cross-trained on operations. In addition, standard operating procedures needed to be established. To ensure success, the employee performance evaluation procedures needed to be re-evaluated and standardized.

Simultaneously, the supply chain team at CCB was suggested to start procurement of raw materials from the recommended distributor for top face of PCBE, side and cross tubes of CTCR. Colorado Extrusion was the recommended supplier for the side walls and base face of PCBE. During this process, o-ther suppliers and distributors were to be evaluated based on lead time and price. However, prior to actual implementation, the equipment and cells had to be established and set up at CCB.

To meet the varying product demand requirements and to eliminate the backlog, it was necessary to establish an "office cell" with a floater. A master scheduler role would be needed in future to maintain forecasts, slot orders for manufacturing, load balancing, in addition to entering customer orders. Management needed to get involved to define this cell and establish appropriate performance parameters to ensure smooth operation at the front end of the process.

CCB needed to incorporate a phased approach to reduce the paint MRP lead time to 1 day. It was important to support this reduction by focusing on identifying obstacles and resolving issues as they came up during implementation.

Alex started thinking about the implementation; this was going to be a big leap for the organization to start the transformation process; he needed to recruit people who would be excited to lead this change. Also, he felt that the project needed to be broken into smaller sub-projects to get buy-in and move the process forward. Alex started drawing up an initial project plan; the details are shown in Table 6.17. It was a formidable challenge for him to guide his team through this cultural transformation from a functional manufacturing structure in a union shop to a lean cellular structure. Successful implementation of this project would lay the groundwork for future transformations at CCB.

7 Meeting Customer Expectations

Classic Case of a U.S. Wooden Cabinet Maker

Ubercraft Cabinets was founded by Jason Smith in 1946 after he finished his duties as an officer in the U.S. Navy. Jason bought a large piece of land in Jasper, Indiana, and constructed his first building using a disassembled, corrugated roof building from his hometown in Evansville. He continued to build additions to the shop, expanding with new buildings in 1963, 1967, 1970, 1975, 1980, 1982, and 1984. In 2001, Ubercraft Cabinets was in a 200,000 square foot facility and was considered a leader in the design, manufacture, and assembly of a wide range of cabinets, consoles, shelves, and other furniture for kitchen, bath, and other rooms in residential and business applications. The facility had full, up-to-date veneer manufacturing, finishing, and installation capabilities. The company was known for its knowledgeable staff with high attention to detail. The products manufactured by Ubercraft were sold through distribution and big retail stores such as Home Depot and Lowes.

Ubercraft manufactured both standard-sized products known as branded products, as well as custom products. Branded products are catalogued products which include standard-sized cabinets, consoles, and shelves with options for different sizes. These products were mostly make-to-stock and Finished Goods Inventory (FGI) was replenished after a product was shipped to the customer. OEM/contract manufacturing products were customized products which include custom cabinets, shelves, enclosure, and other furniture. Contract manufacturing products were made to order according to customer specifications. These specifications were provided by customers either in the form of drawings of the required product or by sharing specifications of design changes required in an existing standard product.

The Order Entry representative managed all orders received by the company. A sales representative would send orders for all branded and repeat contract manufacturing orders to Order Entry group, which, in turn, will reach out to production scheduler and raw material planner to get an estimate on the ship date. Order Entry representative will then enter the information in the ERP system after acknowledgement from sales. The MRP within the company's ERP system was run twice daily. For new contract manufacturing orders, the representative would also include the Engineering representative in discussions. Once details were provided by the concerned engineer, the information was subsequently reviewed by production scheduler and production representative to confirm ship date. The production of both types of products was completed in the facility with a functional layout and used NC saw cutting, NC drills, presses, and manual and automatic assembly stations for cutting, drilling, bending, and assembly of parts, respectively. Painting of the products was carried out in two different areas. Due to their comparatively smaller size and large production quantity, branded products were painted in the main paint area using a conveyor system. Contract manufacturing products, on the other hand, were painted in batch paint booths. Assembly was done along with packing of kits for each specific product.

Contract manufacturing was divided into three groups: Group 100, 200, 300 (Appendix H). Each group was managed by a production supervisor and produced different types of products. Group 300's sales were the highest among the three groups, but Ubercraft was predicting an increase in sales for Group 100 in near future.

All contract manufacturing was done using a make-to-order approach, and production scheduling for contract manufacturing parts involved personnel from several different groups. The three key pieces of information needed for production scheduling were customer need-by date, resource availability, and material availability. By combining information from concerned personnel in sales, production, and procurement at Ubercraft, production orders were generated in different time buckets. On resource availability side, production scheduler together with production managers from the three groups looked at current workload, future workload, part router, and paint shop workload to estimate ship dates. ERP system then looked at the part router to schedule start date of each operation. The production scheduler also used feedback from the purchasing department about raw material availability to finalize ship dates. It used to take 2 days for painting and 2 days for assembly, packing, and shipping in the contract manufacturing area. The order scheduling and entry process is summarized in Figure 7.1.

Justin Sherman, who was the Vice President of Operations at Ubercraft, had recently completed his second year at the company. He had joined Ubercraft after managing Operations at another sheet metal manufacturing firm for more than 15 years. Justin had a strong background in lean manufacturing and implemented several lean projects over the last several years. Last 2 years had not only given him sufficient exposure to many improvement opportunities within the company but he was also more aware of external challenges that the company was facing. He was

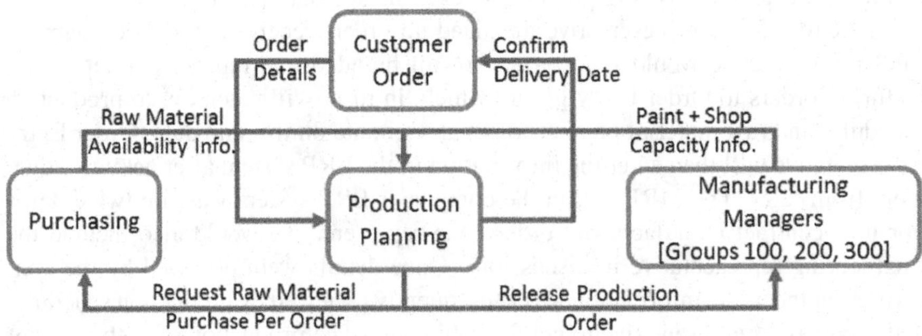

FIGURE 7.1 Overview of order scheduling and entry process.

very keen on making several changes at Ubercraft to better position it to meet customer expectations and retain its market share. According to Justin, Ubercraft had several reasons to make changes. Justin estimated Ubercraft's external lead time (from order entry to product shipping) as 1 week for standard products and 6 weeks for custom contract products. The lead time for contract manufacturing products was long by industry standards, especially for consistent orders. Some of Ubercraft's customers used multiple-sourcing strategy, creating additional pressure on Ubercraft to perform better than the competition. Justin believed that competing on delivery time could help Ubercraft strengthen its relationship with customers and hopefully get a larger share of orders in future. Additionally, most of Ubercraft's customers were in the furniture retail industry which had short cycle times, making Ubercraft's biggest concern as being quick enough to meet customers' needs.

Furthermore, Justin was well aware that longer lead time on contract orders forced Ubercraft's customers to forecast their demand. This was a challenge in an industry with shorter lead time expectation from end customers, ultimately leading to frequent changes in either desired ship date or order mix. Customers frequently called to change order ship dates or quantity leading to extra work for Ubercraft in rescheduling production, as well as raw material procurement.

Justin was also noticing a gradual shift in preference among customers toward custom orders. Ubercraft already had a significant backlog of contract manufactured orders, and Justin was concerned that future increase in orders will be even more difficult to manage. While the margin was higher on contract manufactured products, it also came with the expectation of showing satisfactory response in Engineering and Production-related activities.

Lastly, Justin had recently completed a project that led to a few changes in the manufacturing of branded or standard products, which were made-to-stock. These changes were implemented to reduce lead time so that overall working capital in form of Work-in-Process (WIP) and Finished Goods Inventory (FGI) was reduced. Since the processes for branded products were comparatively better defined and the order pattern was stable, it was an easy first step for them. It also laid the groundwork for making changes in the manufacturing of contract manufactured products.

To analyze and make changes in the contract manufacturing area, Justin called in the same team that had worked on the previous project for standard products. The team consisted of Industrial Engineer Kyle, Finance Manager Mona, and Manufacturing Manager Darryl. Kyle had been with the company for 1.5 years and was hired by Justin after he joined Ubercraft. Kyle was a recent graduate with a degree in Industrial Engineering. Mona had been with Ubercraft for over 5 years and was recently promoted to the finance manager role. She started at Ubercraft as a business analyst after finishing her undergraduate degree in business administration. Lastly, Darryl had been with Ubercraft for over 15 years. He started as a material handler in the warehouse and then gradually made his way up to a machinist role and then as a manufacturing manager. He had been the manufacturing manager for the last 7 years. Justin not only trusted these three based on his interaction with them over the last 2 years but he was also impressed by the work this team had done in the previous project.

During the project kick-off meeting, Mona started by sharing some basic sales-related information with other team members as showed in Figures 7.2 and 7.3. The contract manufacturing sales were close to $30 million, which was 75% of the total company revenue, while the remaining 25% of revenue came from the sale of branded or standard products.

Figure 7.3 shows the breakdown of sales of the three production groups in contract manufacturing products.

During further discussions, Justin discussed details of ongoing issues and opportunities associated with Group 100. Although Group 300 contributed the highest sales in the contract manufacturing category, Justin believed that Group 100 had the most potential for an increase in demand in near future. It was therefore decided to evaluate products within this group to identify project focus. Within this group, it made sense to pick the product(s) with a higher percentage of sales and could benefit the most from a shorter lead time. This would also help get enough visibility on project success within the company creating opportunities for future change. Among various products within this group, Custom Miller Cabinet (CMC) was selected as the project focus due to its largest sales and unsatisfactory lead time of up to 6 weeks.

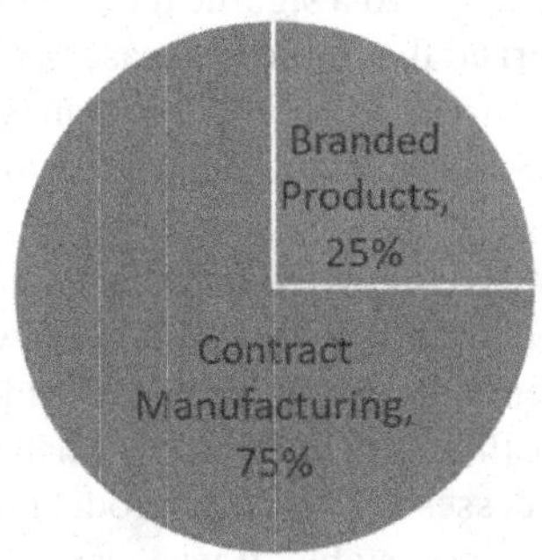

FIGURE 7.2 Breakdown of total sales revenue.

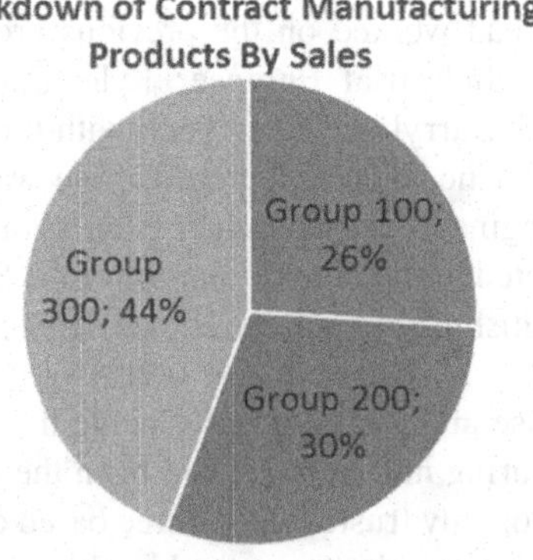

FIGURE 7.3 Breakdown of sales revenue for contract manufacturing products.

Figure 7.4 shows the breakdown by volume of products in Group 100. D-Brand included 12 products for customer Dixie Inc.; L-Brand included 21 products for customer Lincoln Industries; and Misc. includes 7 other products. CMC was 19% of the total volume. Also, the sale of CMC in 2016 was the highest in Group 100. The current quoted lead time of CMC was 6–7 weeks, which was too long according to Ubercraft's customers. The customers for CMC used a multiple sourcing strategy and distributed orders based on lead time quoted by the supplier. Justin, as well as project team members, believed that focusing improvement efforts on reduction of CMC lead time would help create a favorable impact for Ubercraft and help the company create a favorable momentum for similar changes in future.

There were four models of CMC, which differed from each other in terms of size. Although Ubercraft used different product numbers for these four models, their production processes were the same. As shown in Figure 7.5, a Custom Miller Cabinet Unit order started from order entry to saw cutting, drilling and edging, painting, assembly, and then finally ended with shipping. Painting was done in a batch paint booth, which was shared by several other products from the contract manufacturing area. Production process included saw cutting, drilling and edge finishing, assembly, and painting.

Through initial discussions among the team members and additional analysis of available data, it appeared that focusing on the production area alone would provide

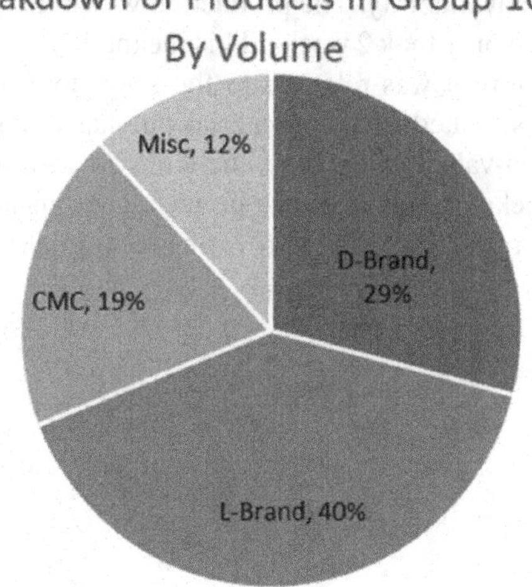

FIGURE 7.4 Breakdown of production volume for Group 100.

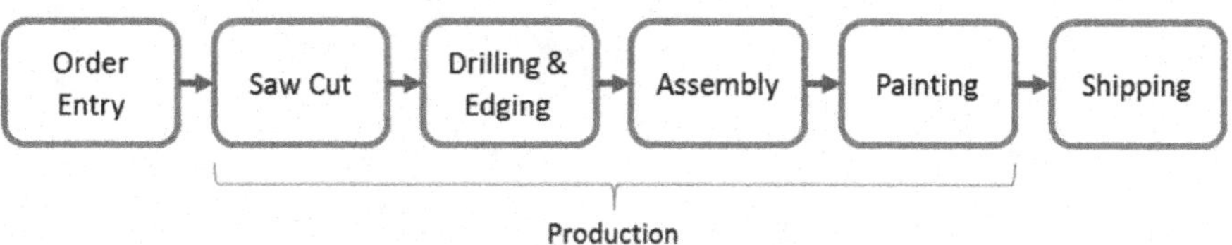

FIGURE 7.5 Order process flow map of CMC.

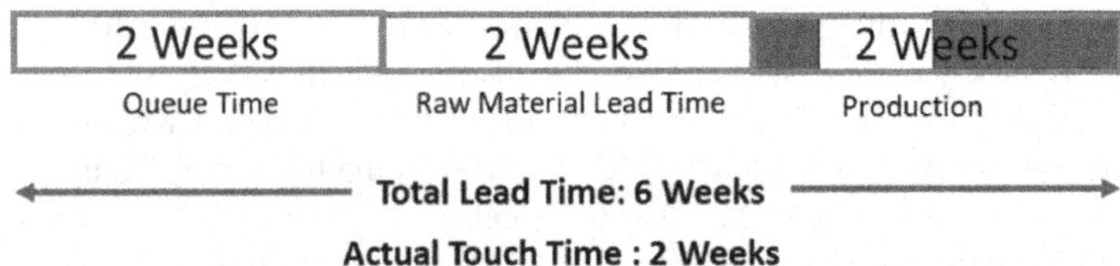

FIGURE 7.6 Value-added versus non-value-added diagram.

enough opportunity to achieve a significant reduction in lead time. As discussed earlier, paint was a bottleneck resource, and it would have made sense to include it in project scope. However, Justin had already hired an external expert to evaluate opportunities for improving efficiency in the paint area. Thus, the final project scope included raw material supply and shop floor operations including cutting, drilling, and finish grinding.

Kyle developed a lead time map for CMC by using his own product knowledge and by analyzing data of CMC jobs finished in last 1 year. The lead time map for CMC in Figure 7.6 shows the breakdown of total lead time, with gray parts indicating value-added time and white parts indicating non-value-added time.

Ubercraft was quoting a 6-week lead time to its CMC customers, who, in turn, had to look in their forecast to estimate order quantity to place order with Ubercraft. The actual ship date was confirmed to the customer after resource planner confirmed resource and material availability. The production which included saw cutting, drilling, and edge finishing only took 2 weeks. This meant that an order waited in a queue for about 4 weeks before it was released to the shop floor. In this 4-week waiting time, 2 weeks were estimated for raw material purchase. Since nearly two-third of the total time was non-value-added time, the team decided to reduce the total lead time by 50% to 3 weeks. In addition, the project scope included developing a new raw material procurement strategy and providing recommendations to improve order placement strategy.

8 Keeping Customers Happy with Shorter Lead Times

Ubercraft's Approach

Once the working team had a consensus on project goal, it made sense to identify key areas of focus within the value chain. This could be either done by data analysis or by using feedback from different key personnel involved in the day-to-day activities. In some cases, getting data was somewhat difficult, but most of the key managers had been in their role for several years. Kyle, Mona, and Darryl decided to organize a brainstorming session with Ubercraft managers and leads from manufacturing, quality, sales, procurement, and order entry. This session was valuable and revealed several key insights. Most important of all, it highlighted that four key contributors to long lead time were frequent order change, resource capacity constraints, raw material procurement, and planning inaccuracies in the painting area. The painting was a bottleneck and many parts were being rejected due to poor surface finish. Justin had already initiated a project to resolve this issue with help from an outside company that had provided the paint equipment, so it was excluded from the project scope.

After defining project goals and three main focus areas, the team developed a work plan for further investigation. All the work started with interviews of relevant company personnel and data collection, followed by observation, root cause analysis, and concluded with specific recommendations for each problem. However, the three main problems discussed above were not independent; therefore, the analysis was performed by considering the interaction among them. The methodology is shown in Figure 8.1.

Table 8.1 summarizes the types of data collected for each specific focus area.

Kyle analyzed the sales history data and found that the actual ship dates of more than 55% orders were later than the scheduled ship dates. Some of the delays were due to a customer's request to reschedule the shipping date, but most of them were due to Ubercraft's inability to meet the requirement. Figure 8.2 shows the record of lead time of each order over the last 2.5 years.

Kyle's analysis of actual lead time data on different orders provided several useful insights. The first observation was related to a large variation in observed lead time. In Figure 8.2, lead time was calculated from the time an order was entered into Ubercraft's system to the time it was shipped to the customer. The actual lead time had a standard deviation of 3.9 weeks with the longest lead time being 18 weeks while the shortest lead time was close to 2 weeks. One of the key reasons for this variation was frequent change in need-by date from customers. Due to shorter lead

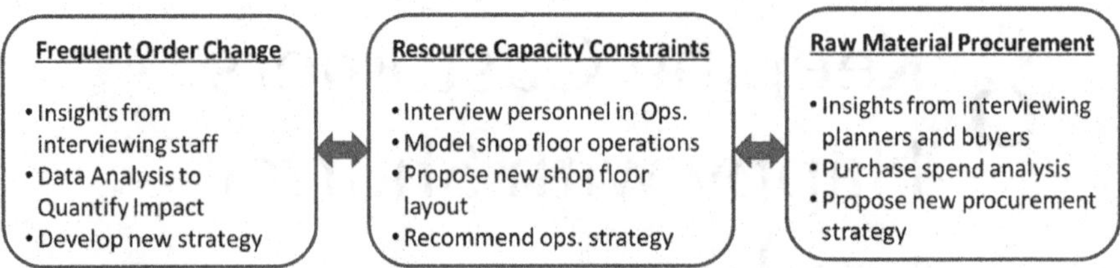

FIGURE 8.1 Project methodology.

TABLE 8.1
Summary of Data Collected for Each Focus Area

Focused Area	Data Collected and Analyzed
Order receipt and entry	Order history
	WIP Data in front of different resources
	Shipment history
Production process	Part routing + setup and run times
	Resource names and availability
	Operators count and availability
	Shop layout
Material procurement	Bill of material
	Material purchase history
	Vendor information

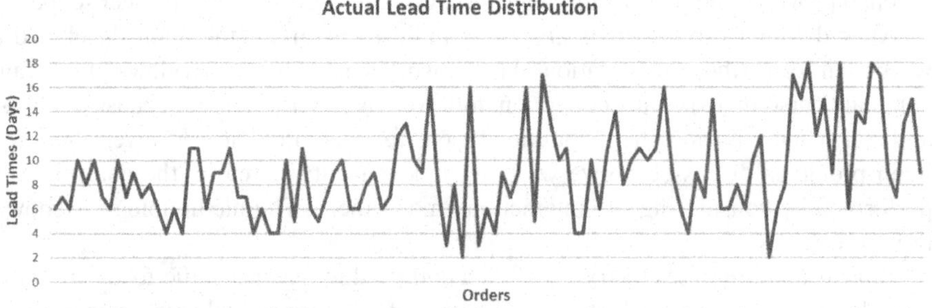

FIGURE 8.2 Lead time for each order over last 30 months.

time expectation from end customers, Ubercraft's own customers in the furniture industry had limited ability to predict their long-term demand accurately. These customers therefore frequently called Ubercraft to change ship date on orders they had placed earlier. Ubercraft usually did not allow order changes 10 working days before ship date, but beyond that period customers called several times to change, update, or sometimes cancel their order. The team had also learned from their brainstorming session that every week, manufacturing planner would schedule multiple meetings with manufacturing manager and sales representative to reschedule production dates based on ship date change requested by customers. Contract manufacturing group estimated that 40% of all orders received were rescheduled by the customers. Furthermore, the planners and managers spend more than 7 hours every week in rescheduling jobs through the shop floor. Out of this, 2 hours were used for rescheduling CMC orders.

One additional observation from the shipment data was that the average lead time was longer than quoted lead time at nearly 9 weeks, which was 50% more than the quoted lead time. This was primarily due to two reasons. First, customer orders were changed frequently leading to incorrect prioritization of orders on a regular basis. Secondly, many customers either pulled in the order date or requested for expediting their orders. This led to rush jobs, which further increased the lead time of regular orders. Darryl was familiar with this issue and summarized how the long response time spiral works in Ubercraft CMC manufacturing in Figure 8.3.

Darryl also highlighted that frequent order changes also resulted in an increased Work-in-Process (WIP) on the shop floor. In past, he tried implementing the practice of setting aside WIP associated with rescheduled order, but it resulted in too many misplaced parts and general confusion on the shop floor. Sending the semi-finished parts back to warehouse also was not possible due to limitations of ERP system as well as challenges associated with keeping an accurate accounting of costs incurred on semi-finished parts.

Production orders were released about 2 weeks before ship date. This production schedule during this 2-week window was frozen to maintain some control over CMC delivery. This means that on an average an order waited in the system anywhere from 6 weeks (quoted lead time) to 8 weeks (actual average lead time). However, to keep customers happy, sometimes they still had to reschedule a few jobs in the 2 weeks window based on customer request.

Darryl further explained to the team that even though the production schedule was largely frozen during the last 2 weeks, Ubercraft still suffered from frequent order outside of this window. This is due to the production strategy adopted by Ubercraft. Due to uncertainty of their own order, Ubercraft's customers divided each order into multiple deliveries. Each delivery request had its own ship date and quantity.

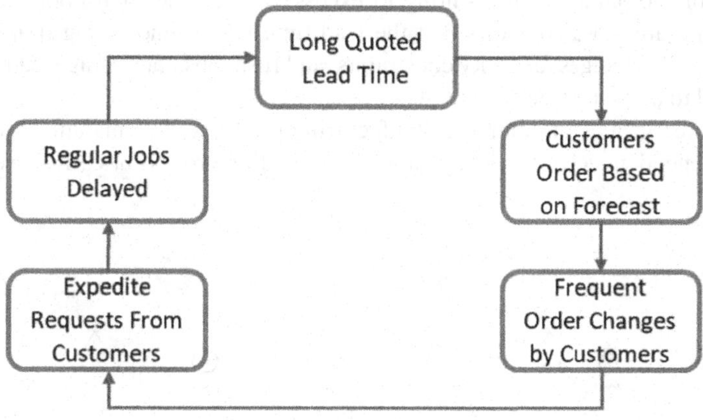

FIGURE 8.3 Long lead time spiral.

This way, customers could get delivery of cabinets more in sync with their own demand while minimizing exposure to working capital tied to inventory. So, for a forecast of 50 cabinets for a given quarter, customers would break the order into five deliveries of 10 cabinets each, with each delivery requested on a different week. As time rolled further and they had a better idea of actual customer order, they would, in turn, call Ubercraft to adjust the delivery dates. This allowed customers to only change a part of their total purchase order to Ubercraft rather than the entire order.

Darryl believed that having orders placed by different customers using the approach of multiple delivery quantities and dates minimized its risk of committing resources in producing a large batch, which may get rescheduled. It further allowed manufacturing to use the WIP in fulfilling the demand for another order that was either rescheduled in or out. Darryl tried to explain this approach using an example as shown in Figure 8.4.

In the figure above, Ubercraft had orders from three different customers with different configurations of the cabinet due in different weeks. Customer A requested for deliveries in week 2, 3, and 4 while customers B and C requested deliveries from weeks 4 to 7. In case of order cancellation by customer A, Ubercraft planner would get together with Darryl and sales representative to discuss possible alternatives to manage inventory in the shop floor. In this example, the resolution was to use order for 7 units to meet the demand for same units from customer B and use order for 12 units to meet the demand for same cabinets from customer C. The sales representative could try to reach out to customer C to take earlier delivery of 12 units, but if they do not agree then these 12 units will likely get finished earlier and wait near shipping area. Since there is no demand for 11 units of third cabinet type from customer A, it will be scheduled to a later date and it waits in the shop floor as WIP.

Furthermore, Ubercraft planner grouped orders to maintain its lot-sizing approach, so orders from different customers or cabinet types were grouped together. This means orders from different delivery dates could be produced together resulting in a few units produced much earlier than its ship date. All these units would sit before the paint area and get prioritized based on firm ship date from the customer. If a customer rescheduled or cancelled the order, then the WIP would wait longer in front of the paint booth resulting in excess WIP on the shop floor. Lastly, the chart in Figure 8.3 also shows that the lead time had an increasing trend. Due to increase order changes, capacity constraints, and rush jobs, the average lead time had increased to almost 9 weeks.

The three groups of contract manufacturing products were manufactured in the three independent cells. To understand the production process of CMCs, we need to

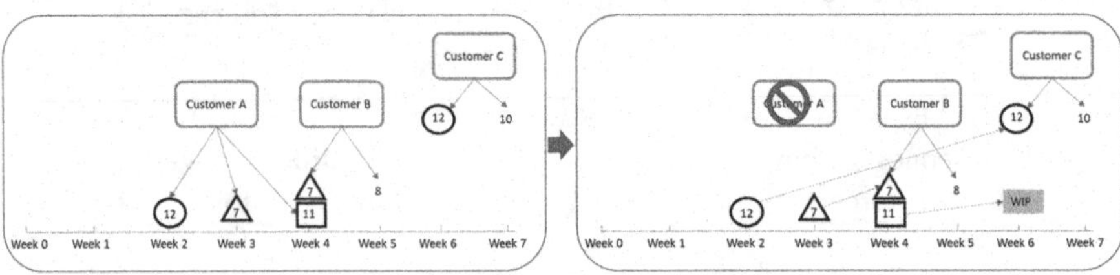

FIGURE 8.4 Ubercraft's strategy in response to order change.

understand the cell of Group 100. Figure 8.5 shows the complete layout of Group 100 cell. There were 44 products in the group. Group 100's cell could be further divided into several small groups namely, CMCs, D-Brand, L-Brand, and Miscellaneous. D-Brand and L-Brand were associated with names of customers who ordered high-volume products of different types. The miscellaneous category included low-volume products, with seven different types of products.

Among the 44 products, 28 were considered high-volume products with four CMCs, 21 L-Brand, and 3 D-Brand. The remaining 16 in Miscellaneous and D-Brand were grouped as low-volume products (Appendix I: Products in Group 100). The 28 high-volume products constituted nearly 88% of the total product volume by number. Table 8.2 shows the quantity and percentage of each product group in both high-volume and low-volume set.

All of the 44 products shared saw and drilling resources including two large saws, two edge milling, two finger joint miter machine, and one drilling machine, whereas the adhesive application and fixture installation were carried out at independent areas

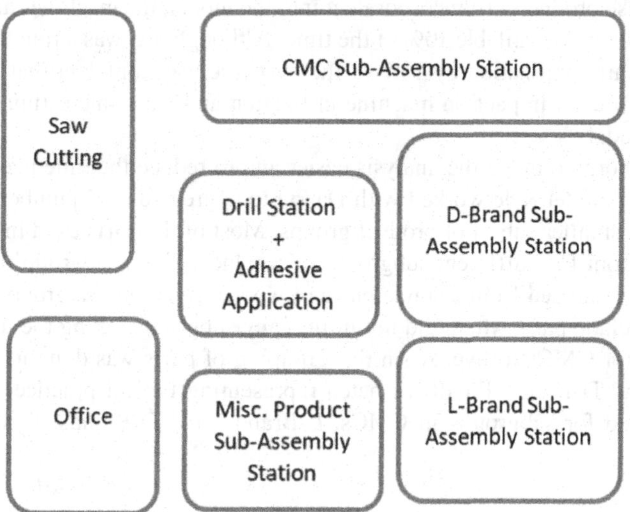

FIGURE 8.5 Shop floor layout of Group 100.

TABLE 8.2
Group 100 Products Breakdown by Volume

		CMC	L-Brand	D-Brand	Misc.	Total
High volume	Count of SKUs	4	21	3		28
	% of units produced	19%	40%	29%		88%
Low volume	Count of SKUs			9	7	16
	% of units produced			5%	7%	12%
Total		19%	40%	34%	7%	100%

for each specific group. As mentioned earlier, CMC shares all the saw and drilling resources with all other 40 products but has a dedicated sub-assembly station.

Since all the products were wooden cabinets, their fabrication processes were very similar. Parts were usually cut on the saw and then drilled on numerically controlled (NC) drill followed by edging on Miter machine. All the parts were then collected and moved to different assembly stations. The complete process on the shop floor is illustrated in Figure 8.6.

Kyle took on the responsibility of building a model of the manufacturing process to simulate the production environment. To build the model he used data such as demand history, Bills of Material (BOMs), routings, machines, and number of operators. Kyle was aware that the router data and BOM data were more accurate for 90% of the high-volume products because the remaining 10% low-volume parts weren't produced as frequently and as such their data was not updated after last updates in shop floor machines. Fortunately, since all the products were wood with similar fabrication processes, his modeling approach was to aggregate all the low-volume products, which accounted for 10% and assumed that 10% of the time was used to produce these products. To incorporate it into the model, the machines and operators were considered unavailable 10% of the time. Although this was a rough approximation of machine utilization by 10% of the parts, he was confident that it would not make a significant impact on machine utilization and lead time estimates obtained from the model.

Furthermore, to make the analysis easier and to reduce the time needed to build a simulation model Kyle worked with Darryl to aggregate total number of products models to a smaller subset of product groups. Most of the parts used in CMCs were generated from four different subgroups of products. Each subgroup represented a set of parts produced from a single piece of wood. One more subgroup of parts was added to include parts which did not fit into any subgroup, taking the total count of subgroups for CMCs to five. A similar grouping of parts was done for products in L-Brand and D-Brand. Finally, a batch representing typical practice by the planner was used for subgroups in CMCs, L-Brands, and D-Brands. Kyle was aware

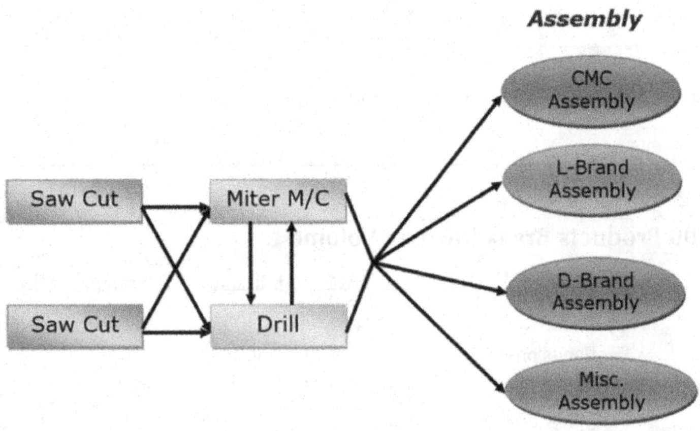

FIGURE 8.6 Group 100 production flow.

that developing a model to simulate shop floor conditions involves keeping a good balance between aggregating products to reduce modeling time, while at the same ensuring it can give relevant insights. So, he allocated sufficient time in thinking through the modeling approach before starting the modeling work.

The shop floor of Group 100 ran two shifts, but not all machines were used during both shifts. Both saws ran in two shifts, whereas Miter edger machine and drill only ran one shift daily. At shift one, dedicated operators worked on saws, drill machine, and Miter machine. In the second shift, Ubercraft had dedicated operators only on one saw cutting. Through discussions with production manager, it was realized that the second saw was run during the second shift roughly 60% of the time.

The simulation results showed that the lead time for CMC starting from production of component parts to the assembly operation was 34 days. Ubercraft used to take 2 days to paint, 1 day to pack, and 1 day to ship out their products. If these three processes were included, the total lead time came out to be 34 + 4 = 36 days or nearly 7 weeks, which was between the quoted lead time of 6 weeks and the average lead time observed from the past shipment history of 9 weeks. Based on further discussions with Ubercraft about the discrepancy between quoted lead time and the lead time observed using simulation model, it became evident that excess WIP in the system was the primary reason Ubercraft was able to meet the 6-week quoted lead time. A lot of this WIP waited in front of paint station which was known to be a bottleneck station due to capacity and quality issues.

The model also showed that other than paint, the CMC production process had bottleneck at drill station with a processing time of 10 days followed by wait at Miter edging station with a processing time of 8 days. Most of the lead time was waiting for the machines to become available and/or waiting for the rest of the lot to be finished. Ubercraft had also estimated that edging processes were highly utilized, as the equipment was shared by all Group 100 products.

The utilization of the Miter edging machine and drill in the model results was close to 100%. High machine utilization with a large variety of parts running through them can significantly inflate the lead time of products. To assure shop floor flexibility to respond to variable demand, it is better to keep machine utilization close to 85%, but existing equipment utilization greatly exceeded these limits. In addition, since both drilling and edging operations were performed manually, labor utilization of the two stations was considerably high. Kyle's simulation model therefore helped highlight key capacity constraints that were significantly impacting the lead time of CMCs. This ultimately helped develop alternatives toward capacity planning and part routing on the shop floor, which are discussed later in team's recommendations.

The BOM for CMCs contained about 45 different purchased components with an estimated lead time of 2 weeks on longest lead time components. From interviews with the raw material buyers and planners, the team realized that majority of these purchased components were standard hardware such as nuts, studs, nails, hinges, and screws. Out of the 45 different purchased components, only 7 components were non-standard hardware. All these components including wood were either vendor managed inventory (VMI) or directly purchased by the buyers. VMI approach was used for high-volume parts consumed by all product groups throughout the company, while only a few of CMC components were directly purchased. VMI inventory was replenished within 2 days after it fell below a predefined trigger point, and the lead time for directly purchased parts ranged from 4 to 7 days. Thus, even though Ubercraft planned as well as quoted around 2-week purchased component availability, its actual lead time for all purchased components was 7 days or less. There were a few reasons for padding this raw material lead time based on previous experiences of material shortages.

The team had learned from its recent projects in other areas of the company that one of the main reasons for material shortages was use of standard VMI items by all product groups. Items such as nuts, nails, and screws were used by both branded products with higher usage volume as well as custom products. Since the forecast for branded products was not accurate, many times a larger than planned batch of a branded product would consume all the VMI inventory. The reorder points (ROPs) as well as order quantities were based on intuition and past experiences rather than based on historical analysis of component usage and desired service levels. The supplier would usually try to replenish the VMI quickly, but Ubercraft used to run out of one or more items more than once a month.

Secondly, the directly managed parts specific to CMC were ordered based on order confirmation. These parts were non-standard and relatively more expensive, so the intent was to minimize working capital tied to these inventory items. The company's ERP system would prompt the buyer to purchase material, but it was based on the required ship date and lead time of different configurations of CMCs. By analyzing the system lead times Kyle observed that it was not accurate and was saved as default 10 days for all products. This highlighted the need to maintain key item attributes in the system. The planners and buyers were expected to rely on system information to order raw material, but the data leading to the calculation of those order dates was never updated or maintained. Over time they had lost faith in the order date suggested by the system and had started ordering material based on their own estimate, which often resulted in mistakes.

Finally, a regular practice of shuffling orders had compounded this material shortage problem. Jobs never really started when they were planned in the system. The manufacturing manager planned production based on ship date, bottleneck in front of different machines, and availability of buffer inventory in form of WIP at different workstations. Many times this insufficient sync between operations and material planning led to material shortages by actual production start date.

During a status update meeting, Justin pointed out that these three problems identified by the team were commonly observed among manufacturing companies. The leadership team picks an ERP system for the company and insists on planners and managers to use it to plan their activities. Lack of emphasis on Master Data Management (MDM) over time leads to a lack of integrity in data, which ultimately drives incorrect material and manufacturing planning. Over time planners and buyers in different areas start using their own estimates to make up for these errors. Due to lack of coordination between these different estimates, the problem either becomes worse or additional padding in capacity and lead time by all groups leads to an inefficient shop floor with low utilization and longer lead time. In the case of Ubercraft, the planner compensated for these stock-outs by adding extra time to order material. The non-VMI was therefore ordered based on a padded lead time

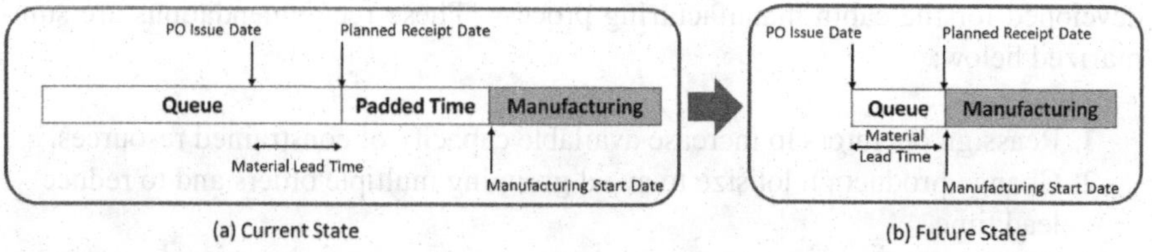

FIGURE 8.7 Procurement of non-VMI materials.

of 2–3 days beyond the quoted lead time from the supplier as shown in Figure 8.7a. Justin agreed with the team that this was an important issue and promised to follow up immediately on the recommendations made by the team.

With an estimated 50% reduction in total lead time in the future, the team knew that the existing raw material procurement strategy would not work well. If Ubercraft desired to significantly reduce its lead time, it would also have to plan for better coordination between manufacturing and material planning. With a shorter or no padding in manufacturing (resulting from shorter queue time in front of machines), the material would also be expected to be available and that too on a shorter notice. Figure 8.7b shows the alternative approach to material planning with shorter total lead time.

During the discussions related to rescheduling, Justin also encouraged the team to explore the possibility of automating several repetitive tasks related to the production of standard orders, so personnel time could be freed up to focus either on improvement activities or non-standard order scheduling. This would ultimately reduce the lead time through the shop. The team reached out to an outside expert to learn more about Robotic Process Automation (RPA). They learned that the basic process behind RPA involves personnel performing their day-to-day tasks while the system runs in the background learning all inputs over some duration depending on task complexity. After the training period for the system is over, the system would perform the task and there would be a control tower to quality check the output of the task. As mistakes get corrected through the control tower, the system's accuracy to complete the task would improve. The control tower tasks would be maintained by the best performing employee in the department who would monitor and improve the performance of the system. The ability to use RPA would allow employees to switch their focus to more thoughtful and meaningful work while also eliminating data-entry errors that can slow processing times, compliance, and the overall customer experience. Two personnel were dedicated in the shop for manufacturing scheduling and a lot of their time was used up in coordinating order changes from customers. The time-phased report output from ERP was studied, and it was realized that fixed routines could be applied to schedule job released for the standard jobs. This would also reduce or eliminate the time spent by different employees in rescheduling meetings. Non-standard item routing, however, still needed manufacturing engineering interaction for routing creation/modification for appropriate scheduling.

The team's final recommendations to Justin and other leadership team of Ubercraft incorporated insights obtained from the analysis of raw material procurement, order management, order rescheduling approach, as well as simulation model, developed for the cabin manufacturing process. These recommendations are summarized below:

1. Reassign operators to increase available capacity of constrained resources.
2. Change production lot size to avoid grouping multiple orders and to reduce lead time.
3. The non-VMI parts need changes in procurement strategy to better meet the needs of shorter lead time.
4. Take a holistic approach at planning demand for vendor managed raw material inventory.
5. Explain the benefits of Ubercraft's shorter lead time on CMCs to customers to reduce future order changes and encourage ordering for smaller batches based on actual demand.
6. Establish a new process for MDM in ERP system.
7. Implement RPA to automate manufacturing scheduling.

1. **Reassign Operators to Increase Available Capacity of Constrained Resources**: The manufacturing process model showed that the drill and edging machines were over-utilized due to operator unavailability in the second shift. On the other hand, the saw was occasionally operated on both shifts, where the machine utilization was less than 70%. To increase the available capacity on edging machine, the team suggested to cross-train saw operator on edging machine, so that both saw and edging machine were available in second shift.

   Furthermore, the assemblers of L-Brand, D-Brand, and CMCs often had low utilization, thereby making them available to run the drill machine. It was therefore recommended to give priority to the drill operation by releasing the assemblers partially from assembling their respective products. This would increase the available capacity of the drill machine while at the same time slightly increasing utilization of the assembly station. There were a couple of different ways in which the shop floor supervisor could implement these recommendations. The first approach is to schedule a fixed amount of time for selected operators or assemblers on the drilling or edging machine. This makes scheduling easier for the shop floor supervisor, but the capacity allocation will not be in sync with daily demand. A better approach that Ubercraft decided to take was for the scheduling supervisor to generate a schedule for operators on a weekly basis. Using this approach Ubercraft could allocate resources as and where capacity was needed for the week.
2. **Change Production Lot Size**: Team also suggested to reduce the batch size of CMCs from 20 to 10, so parts in smaller batches could move to the next station quickly. Running large batches is a standard practice in manufacturing, as it saves time on setup. In case of CMCs, Kyle's simulation model showed that cutting batch size in half for CMCs would significantly reduce CMC lead time without constraining the shop floor resources significantly. The assemblers were already dedicated to CMCs, so there was no need to change setup from one product to another. Also, additional capacity was

made available through cross-training of operators for drilling and edging machine. Furthermore, analysis of shipping history over the last 12 months showed that majority of shipment to meet customer delivery request was associated with order quantities of 10 or less. This is shown in Figure 8.8.

Figure 8.9 summarizes the distribution of shipped quantity to meet customer delivery requests. The histogram shows that more than 73% of quantities shipped to meet customer delivery requests were for 10 units or less. This recommendation was also in line with what customers wanted. With a reduction in quoted lead time to customers, there was no need for them to issue multiple deliveries ahead of time and then reschedule the deliveries closer to the due date. The simulation model showed that Ubercraft would see its lead time reduce by 67% from 36 days to 12 days once all the

FIGURE 8.8 CMC shipped quantity history over 12 months.

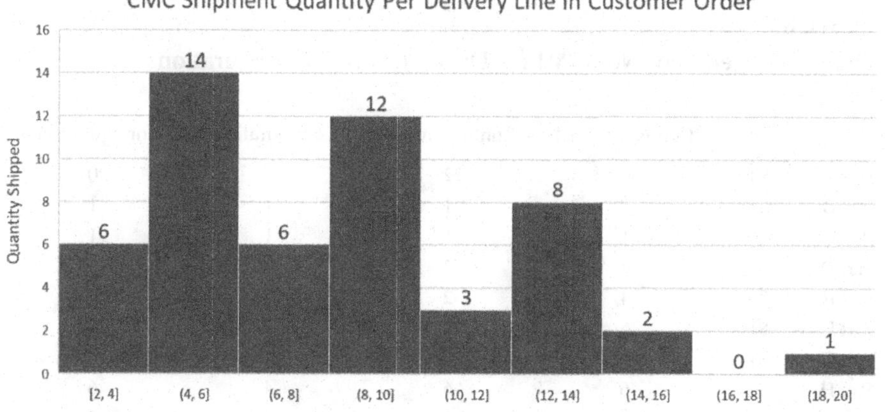

FIGURE 8.9 Histogram of CMC shipped quantity over 12 months.

recommendations on the shop floor are implemented. The shorter lead time was expected to create a greater flexibility to meet customer requirements.

3. **The Non-VMI Parts Need Changes in Procurement Strategy to Better Meet the Needs of Shorter Lead Time**: A reduction in lead time by 67% to 12 days would necessitate a change in current purchase-to-order strategy with up to 10-day lead time for non-VMI parts. Therefore, the team made two key recommendations to address this issue. First was to negotiate shorter lead times with suppliers of these parts. Vendors for non-VMI parts were all located in Indianapolis, which was only 2.5 hours from the Ubercraft facility. The purchasing manager was also convinced based on his interactions with the suppliers that it was possible to reduce lead time for these parts from 10 days to 3 days. The second recommendation for non-VMI parts was to establish a Reorder Point (ROP) system in the company's ERP. This would help bring further trust in material availability of CMC parts, which in past had seen stock-outs due to demand fluctuations.

   Mona compiled the list of non-SMI material and calculated the ROP using the formula listed below:

$$\text{ROP} = \text{Average CMC demand in replenishment period} + \text{Safety stock}$$

$$= d \times LT + Z \times S_d \times \sqrt{LT}$$

   Where, d is demand in the replenishment period, LT is supplier lead time, Z is constant associated with the service level, and S_d is the standard deviation of weekly demand.

   To calculate ROP for each component, she broke down the weekly demand for last 48 weeks for CMCs into different configurations. This is because the quantity of non-VMI parts needed for each configuration was different as shown in Table 8.3.

   The weekly demand for each configuration is shown in Appendix I. By combining the demand data of each configuration with the quantity

TABLE 8.3
Quantity Needed of Non-VMI for Different CMC Configurations

	Unit Cost	Configuration 1	Configuration 2	Configuration 3	Configuration 4
Part A	$40	30	22	18	20
Part B	$12	1	1	1	1
Part C	$15	1	1	1	1
Part D	$3	4	2	4	2
Part E	$3	4	2	2	4
Part F	$10	6	2	6	2
Part G	$10	6	2	4	4
Part H	$10	6	6	4	6
Part I	$7	4	2	4	2

TABLE 8.4
Re-order Points for Non-VMI Materials

	Part A	Part B	Part C	Part D	Part E	Part F	Part G	Part H	Part I
D	208.5	8.5	8.5	27.3	27.8	37.8	38.3	48.1	27.4
LT	0.5	0.5	0.5	0.5	0.5	0.5	0.5	0.5	0.5
Std. Dev.	100.9	4	4	13.2	13.7	20.6	19.6	22.8	14
Z (95%)	1.65								
Z (99%)	2.33								
ROP 95%	222	9	9	29	30	43	42	51	30
ROP 99%	270	11	11	35	36	53	51	62	37
Unit cost	$40	$12	$15	$3	$3	$10	$10	$10	$7
Inventory cost for 95%	$8,880	$108	$135	$87	$90	$430	$420	$510	$210
Inventory cost for 99%	$10,800	$132	$165	$105	$108	$530	$510	$620	$259

needed for each configuration, she calculated the ROP for each part for both 95% and 99% service level as shown in Table 8.4. Her calculations also showed that total one-time inventory cost for maintaining the inventory would be nearly $11K for 95% service level and $13K for 99% service level. Based on discussion with Justin, it was decided to maintain a ROP for 95% service level.

Mona also estimated that the excess inventory would result in an annual inventory holding cost of $2200 at an annual holding rate of 20% for Ubercraft. The detailed weekly demand for each component is shown in Appendix J.

4. **Take a holistic Approach at Planning Demand for Vendor Managed Raw Material Inventory**: Since the vendor managed inventory of several raw material part numbers were shared by multiple product groups with high usage volume as well as high variation in demand, many of these part numbers had shortages in recent past. This concern was discussed with the larger group of people as well as vendors to come to a resolution. One of the suggestions that the team made to the larger group was to collect annual forecast as well as weekly demand variations to calculate safety stock, ROP, and order lot size in collaboration with the vendors. This would also allow vendors to stock inventory appropriately at their end. Furthermore, the team also recommended the purchasing team to renegotiate its supplier agreement with stock-out cost-sharing option added to it.
5. **Explain the Benefits of Ubercraft's Shorter Lead Time on CMCs to Customers to Reduce Future Order Changes and Encourage Ordering for Smaller Batches Based on Actual Demand**: Ubercraft's ability to implement smaller lot size also depended on customers placing orders in

smaller quantities. As discussed earlier, Ubercraft's long lead time and overall lack of clarity on the forecast from end customers had led to the current practice of customers placing orders in large batches with multiple delivery dates and then frequently changing those delivery dates or quantities or both. The team, therefore, recommended that the company inform its customers about changes implemented on the shop floor and the resulting reduction in its lead time. This would not only give Ubercraft the opportunity to differentiate itself from its competitors in eyes of the customers but also help Ubercraft make its case of requesting smaller but firm order quantities. A shorter lead time will also eliminate the need for Ubercraft's customers to put additional work in coordinating receipt of different orders.

6. **Establish a New Process for MDM in ERP System**: MDM is often overlooked in many organizations. Although it is a shared responsibility, many organizations have formalized regular review and cleanup process either by hiring MDM analysts or by implementing formal review processes. In the case of Ubercraft, MDM was not formally defined and over time key item attributes such as item lead times, safety stocks, vendor names, etc., were not reliable enough for use by anyone. Buyers and planners often used their product knowledge and experience to override incorrect inputs from the system. To correct this issue, the team recommended implementing a monthly review process of key item attributes. The team also worked with representatives of Manufacturing, Supply chain, Sales, and Engineering to develop guidelines for review of key parameters. These key parameters included lead time for make and buy items, lot sizes, safety stock, primary vendor name, vendor location details, estimated wait time for each resource, machine availability and shift details, router review for allocating correct resources, setup times and run times, etc.
7. Implement RPA to automate manufacturing scheduling: The team believed that RPA would help Ubercraft streamline the job scheduling function. The external company consulted for this would help automated standard jobs scheduling in various functional areas of the shop through a programmed "bot." The program would learn the manual job scheduling routine for standard jobs as it got inserted into the schedule, through a control tower function. Streamlining this scheduling function would allow smoother flow through the shop while reallocating resources to focus on bottleneck areas for improvement purposes. The external company also worked with the project team to prepare a cost analysis tool for management to evaluate the financial impact of the RPA implementation. The analysis included the following input parameters (Table 8.5):

Tasks the RPA bot would automate: 2
Number of employees currently perform the task(s): 3
Percentage of the employee's daily time spent on these tasks: 40%
Employee average salary: $55,000/year

The calculated payback for this investment was estimated to be 1.37 years.

To estimate the impact of lead time reduction on the shop floor WIP, the output from the simulation model was combined with Little's Law. Little's Law's states that in a stable system, the long-term average number of parts in the system (WIP) is equal to the multiplication of throughput rate and average lead time of the system. Using the observed throughput of 10.5 units per day for CMCs, the impact on WIP from shorter lead time is summarized in Table 8.6.

While the reduction of WIP from lead time reduction looks dramatic, Darryl and the rest of the team believed a stepwise approach would allow them to validate results and adjust as needed along the way. The team's recommendation was to make changes in resource allocation immediately, but the reduction in lot sizes would happen in steps. This would allow shop floor supervisors to assess the effectiveness of changes and make any adjustments if needed. This stepwise approach to lot-size reduction would also require a stepwise update of queue time in the system, which would ultimately lead to a gradual decrease in WIP.

Mona calculated that the reduction in WIP of CMCs by 67% would help Ubercraft realize a one-time saving of nearly $6300 assuming a 20% holding cost. This was calculated by estimating WIP at half the cost of a finished unit and multiplying WIP value to reduction in WIP count and the holding rate. The sales group at Ubercraft was also very excited about the prospects of a drastic reduction in quoted lead time from 6 to 2 weeks and indicated that they would increase the overall sales target for CMCs by 5% next year if manufacturing could successfully deliver products in this shorter lead time. This gain in sales will come from a combination of increased order from existing customers as well as through acquisition of new customers. The team summarized the overall cost-benefit analysis of the project in Table 8.7.

TABLE 8.5
Cost-Benefit Analysis and Payback Calculation of RPA Implementation

	Year 0	Year 1	Year 2	Year 3	Year 4	Year 5	Total
Current employee costs ('000)		$165	$165	$165	$165	$165	$825
Employee hours/year that will be automated with RPA		2400	2400	2400	2400	2400	
Employee cost/year that will be automated with RPA ('000)	$0	$66	$66	$66	$66	$66	$330
Cost to implement RPA bot ('000)	$70	$15	$15	$15	$15	$15	$145
Net savings ('000)	($70)	$51	$51	$51	$51	$51	
Cumulative saving ('000)	($70)	($19)	$32	$83	$134	$185	

TABLE 8.6
Impact on WIP from Shorter Lead Time

	Throughput (Pcs/day)	Lead Time (Day)	WIP (Pcs)
Current	10.5	36	378
Future	10.5	12	126

TABLE 8.7
Cost-Benefit Analysis for the Project

	Year 0	Year 1	Year 2	Year 3	Year 4	Year 5
WIP reduction ('000)			$6.3			
RPA implementation ('000)	$70	$51	$51	$51	$51	$51
Profit from increased sales ('000)		$79	$79	$79	$79	$79
Non-VMI material holding cost ('000)	$2.2	$2.2	$2.2	$2.2	$2.2	$2.2
Layout changes + operator training ('000)	$5					
Net saving ('000)	$77	$134	$128	$128	$128	$128

The primary expense for Ubercraft in implementing these recommendation results from use of the RPA system, which is estimated to have a payback of 17 months. Only other recurring expense results from holding correct amount of non-VMI material to avoid stock-outs. The overall project recommendations were expected to have a payback of less than a year.

The team also developed a detailed implementation plan that primarily focused on shop floor rearrangement, RPA implementation, raw material supply management, and workforce management to ensure better buy-in to these changes. The team recommended Justin to start the implementation by having a formal kick-off event where the details of the proposals will be shared with all company employees. The team also suggested outlining details of future state once these recommendations were implemented. A strong support base to embrace these changes was important to build the momentum for change, especially for shop floor operators who would be cross-trained and in future will work more like a member of the CMC team. The immediate changes on the shop floor would begin with layout changes, task allocation, and operator cross-training, which was expected to take 2 weeks. The team also developed a performance measurement system for CMC staff that included overall CMC group's performance in terms of quality, lead time targets, and shipments.

The RPA implementation was a part of central initiative as it would have impacted multiple product groups. Justin believed that it will be able to reassign available free time of concerned personnel after RPA implementation to other areas where they were looking for additional resources. One such area where these personnel would be needed is analysis and update of MDM attributes. These reviews would be part of regular business practice, but the first few rounds of changes were expected to be more time intensive. The RPA system implementation and future recurring support would need outside consultants, but Justin also designated two super users from the same group of three employees primarily involved in order management who would review the output for training the system in early stages.

Simultaneously, based on the team's recommendation, Justin also assigned the task to implement recommendations related to both VMI and non-VMI raw material to one of the leads in the supply chain group. For suppliers of non-VMI material, the leads had to either negotiate shorter lead time or switch to alternate vendors or distributors that could meet the lead time requirements with minimal or no significant impact on pricing. The leads also planned to make changes in ROP right after the negotiations. The current estimates on ROPs were based on variations in ordering pattern under current practices. However, it is expected that these variations would probably decrease as key customers get used to ordering in consistent and smaller batches. However, periodic MDM review process would ensure that such changes are appropriately updated in the ERP system.

The initiative lot-size reduction needed careful planning, since a smaller lot is not preferred by operators. Darryl, therefore, decided to follow the approach of reducing the batch size from 20 to 10 in three steps of 3, 3, and 4 over a 6-week period. Showing the impact of these changes on overall lead time was important, so he wanted to be actively involved during this period to make sure resources are appropriately allocated and any other bottlenecks are addressed promptly to ensure success. It was also important to ensure that the impact on lead time from these changes was communicated to all employees of Group 100 to generate confidence in the new production approach.

This project had a special importance for Darryl, who had been the manufacturing manager at Ubercraft for many years. He was a firm believer in efficiency-based measures that in many ways required running larger batches to capitalize on setup times. When he started at Ubercraft, the product mix was smaller and the volumes were higher, so these ideas worked great. However, with time, as the product mix started to increase and Ubercraft started to lose market to low-cost players, he was beginning to realize that many of the established processes needed a second look. He was grateful for being selected by the company to undergo a lead implementation workshop, which allowed him to better understand the shop floor dynamics. In his previous project, which was also sponsored by Justin, he got the opportunity to test out a few of his newly acquired lean-based thinking. But this time he was working in the CMC area where he spent most of his time while at Ubercraft. Many of the changes that the team had recommended on the shop floor were initially put in place either by Darryl or his predecessors. So he was excited when this project was finalized, since it allowed him to correct a few of the mistakes that he had unknowingly implemented in the CMC area.

Lastly, the team also recommended to Justin and other members of the leadership team that once the changes made on shop floor start showing consistently shorter lead time, Ubercraft representatives should get in touch with customers to highlight these changes either through facility tours or by meeting company representatives. These discussions with customers should also include the possibility of implementing a new ordering strategy of smaller quantity under a shorter lead time of 2 weeks.

9 Lessons Learned

One of the objectives of this book is to walk readers through problem solving in a real-world scenario. Problems are usually complex due to the long history of a company, overlap of different products with existing resources, and issues related to introducing change among the workforce. The learning curve while working on these projects is steep. The discussion in this chapter is aimed toward getting readers acquainted with non-technical aspects of managing a large-scale improvement project.

One of the key approaches to note in all the three cases discussed in this book is a structure-based problem-solving approach. All cases begin with an understanding of the problem faced by the company. This is a key step, but it can often lead to incorrect interpretation if the problem is not understood completely. All three cases usually begin with company overview followed by general feedback provided by the company's CEO or VP about changes outside the company that have made them less competitive and lose market share. In most instances, this preliminary information from the company's head was followed by discussions with leaders and mid-level managers about what they have done to respond to these challenges. The insight shared by the company's leaders is then combined with analysis of their performance data to assess the magnitude of their problem. Once the case has been established, it is followed by a more in-depth analysis of manufacturing processes and products to develop the scope of the project.

The leadership traits that were common among the project sponsors in all three companies were a communication of urgency and vision for change. Once the goal and scope of the project were identified, they all made sure to communicate the reason for initiating the project and reiterated the challenges that the company was facing. They ensured that key managers in the company were involved during different stages of the project to further emphasize the urgency and importance of the initiative. Additionally, they all indicated the need to get a quick win to make a larger part of the workforce start believing in the idea.

For a leader evaluating external threat from competition, it is important to realize that success will result from differentiating the company from its competition. A company can differentiate itself in terms of cost or in terms of other features that result from a combination of efforts from sales, engineering, marketing, and operations. Unless the product being sold is unique and has no comparable alternative in the market, a company must be clear on whether it wants to compete in terms of cost or other features of product or customer experience. In cases discussed in this book, the leadership in operations took initiatives to differentiate itself from the low-cost competition by focusing on lead time. Depending on the product being sold, differentiation could also be achieved by focusing on marketing, sales efforts, distribution network, or unique product features. However, the differentiating initiatives by different functions within an organization must also complement each other. For example, an effort by sales to attract customers through discounts or price breaks may clash with a parallel effort by engineering to differentiate product through unique features or customization. Or an initiative by sales to gain customers by offering customized products could clash with initiatives by operations to focus on high volume and low variety. For leaders of most organizations in the developed world competing with companies in low-cost countries, the key will be to figure out what combination of features will make customers value their offering over lower cost.

Within the developed world, one of the observations among companies feeling the pressure from low-cost competition is to try to position themselves to also compete on cost. Many times managers approach this objective by outsourcing individual operations or manufacturing of subcomponents to an outside supplier with lower cost. In a time of dwindling sales when the overall utilization of existing resources is already low, this approach could further exacerbate the situation by creating a death spiral. This happens when the cost of the product increases due to higher overhead being spread over an increasingly smaller number of products built in-house. During our experience over the last 15 years, we have seen many small companies falling into this pitfall. We believe that a company in a developed world losing market share due to low-cost competition should respond by taking a more holistic approach toward changing existing business strategy rather than just trying to lower costs.

There are some basic differences in operating cost between a company in the United States or Europe and a company operating in China, India, or Mexico, which makes it more difficult for companies in the developed world to compete. In the past few decades, outsourcing helped capitalize on these lower operating structures in low-cost countries, but there are a few key changes in recent years that may be contributing to a shift in this approach. First one is an increased preference among customers for customized product, which is a challenge for the outsourcing model that relies on high volume and low mix of products. The outsourcing model also has an inherent long lead time due to associated shipping time. Most companies overcome this approach by strategically holding inventory at different stages of the value chain, and there is holding cost associated with that inventory. As the mix of product increases and the size of demand for any product type decreases, the holding cost may make the outsourcing model less attractive. Additionally, recent geopolitical changes and government's incentives or tariffs with the intent of reviving manufacturing back to the developed world has created an additional incentive for companies to reconsider the outsourcing model. Lastly, the vulnerability of existing supply chain structure during a global pandemic like COVID-19 has further highlighted the need to review existing manufacturing approaches to make the supply chain more resilient.

Supply chains need to build effective risk management models when companies decide to source from low-cost countries. From a supply chain strategy standpoint, the first thing companies need to think about is the management of spend. Getting a clear idea about how much they are spending over a period and with which supplier and commodity are vital. In order to cut costs and add value, companies can focus on demand management, real-time data analytics, partial outsourcing, and volume bundling. Most companies maintain data on on-time performance and quality performance of suppliers. Creating a multiplicative measure which weighs these two factors will enable companies to create a single measure of supplier performance. The standardized data will allow comparing several vendors across different commodities before segmenting into categories for consolidation purposes.

Understanding the financial position of the suppliers and how many SKUs are single-sourced through them is also critical to developing the risk model. Web-based services like Rapid Ratings and Dunn & Bradstreet in the United States provide financial insights for privately held companies to extract this information.

The availability of data and continuous monitoring will enable companies to configure sourcing strategy as the nature of demand for their products and services changes. These efforts could be prioritized by rating each vendor based on their impact on business and the probability of risk to business from those vendors. The vendors that fall into a higher risk category and are also critical to business need continuous and extensive management and should even be evaluated for possible replacement. On the other hand, a company could use a hands-off approach with vendors that pose a lower risk and are less critical to the business.

The next area of supply chain strategy is to focus around working jointly with the supplier to establish synergies that will be advantageous to both. The scope of this strategy can range from joint demand and capacity planning to closely merging the entire value chain. We have seen examples where most Japanese automotive OEMs have worked very closely with their suppliers by setting expectations for cost and quality. Supplier performance for these suppliers was very closely correlated to the desires expressed by the OEMs.

For an employee who has been assigned the responsibility of implementing the improvement project, the task comes with many unique challenges that an external consultant probably will not encounter. An employee obviously has the advantage of familiarity with the company's products, processes, and people, but the task of bringing in change is difficult for two key reasons. First, there are some limitations to their ability to think outside the box or bring new ideas as it is usually acquired by working across different industries and companies. But in most instances a good understanding of lean concepts, operations management, and supply chain management acquired through education and/or experience can help overcome this hurdle. For example, Darryl at Ubercraft was able to combine his manufacturing expertise acquired through several years of experience with newly learned concepts on lean manufacturing to develop a better understanding of manufacturing dynamics. This allowed him to better analyze existing processes and come out with ideas for improvement. The second challenge for employees working on the improvement project is related to interacting with people they know for a while. This is less of an obstacle for a consultant who is temporarily working on a project. For the employee, the key is to maintain constant and clear communication with the project sponsors as well as relevant employees and managers in the area or group that will be impacted by the project. A good practice among project managers or consultants working with different stakeholders is to ensure "Who-What-When" is understood by all attendees at the end of each meeting. A standardized template covering these three areas not only makes follow up meetings easier, but also helps bring more accountability among the team members for completing tasks on time.

It is also important to learn to deal with changes while doing the project. The project scope sometimes changes a few weeks into the project, which could delay the progress. But there is little that one can do in such situation and the only way out is to work harder to catch up. These projects also require team members to think outside of the box. Many times, it is difficult to get data to develop an insight or conduct root cause analysis. In such instances, the project team is expected to find an alternative approach that could help get to the end objective. This is one of the most common challenges that one faces when working on a project.

Data analysis is the backbone of most process improvement projects. However, most small to mid-sized companies do not have resources in place to collect different types of shop floor data, especially those related to lead time measurement. In all three projects discussed in this book, companies were not directly measuring many of their key non-financial performance metrics including product lead times. In case of lead time measurement the alternate approach used in those projects was to use traveler sheets in which all personnel performing any task on a job were asked to note down the date, time taken, and task performed. A collection of such traveler sheets from multiple jobs over several weeks helped build lead time map showing touch time for different steps in the manufacturing process. In companies that use an ERP system to track traditional efficiency and utilization measurements, a similar lead time map could be developed by using transaction date and time for completing each router step.

Every change of management effort is a project. In most organizations, there will be a collection of projects that need to be synchronized to achieve the result. The project leader needs to identify the bottlenecks and schedule other work based on the schedule of the bottleneck. We saw this in the context of the CCB Products Corporation case where Alex Stolz chose a product line to work on but had multiple dependencies on other functional areas. Before Alex could start working on creating the cell in the shop floor, material testing needed to be conducted which pushed the project further upstream to start with creating an office cell that involved Engineering. The order entry process was the principal bottleneck in the process of creation of the cell. Appropriate buffers need to be placed and watched carefully to prevent work stoppage in case of necessary changes. It is also important to understand that all buffers for managing an individual project, whether they are project buffers, feeding buffers, or resource buffers, protect the bottleneck and thus overall performance of an organization. Overzealous project managers try to control the performance of each step and possibly buffer each step with a lot of inbuilt safety time. While the only thing that counts is the performance of the project as a whole and not the individual steps, the inbuilt safety time in projects feeds a whole different mentality of doing last-minute work since personnel become relaxed about having additional time to do the work. The true answer lies in proper management of the bottlenecks in the project and scheduling it appropriately with the required buffers in place.

In recent years, more sophisticated techniques for data collection like the use of IoT have allowed building sophisticated manufacturing simulation models and the application of machine learning techniques, which allow quicker scenario analysis. The output from the scenario analysis needs to be shared with business leaders to drive the decision and investment processes. Quick decision making will allow organizations to be agile and make strategic adjustments as business conditions change. Another potential area for manufacturing organizations to investigate is Robotic Process Automation (RPA). RPA has the potential to simplify routine tasks and improve their accuracy. Taking the routine tasks out of the employee's bucket allows them to focus on meaningful tasks and develop efficient problem-solving skills. Companies must think about strategically investing in these resources to stay competitive in the market. Traditionally, armies of consultants were needed to set up and keep the systems tuned to produce the required output, but the shift toward the development of Auto Machine Learning (AutoML) tools is creating a new way of thinking about data science. Developing a team internally that understands the business domain, has basic data science skills and willingness to learn the AutoML tools will be the talent pool companies need to develop. Internal expertise can make important adjustments to the models as business conditions change, and that is going to be important to maintaining the competitive edge and reduce the dependency on sparse technical talent or external consulting resources.

It is important to listen to the workers and managers on the shop floor in addition to data analysis. Using just one of these two approaches could lead to incorrect conclusions. A good check to verify the conclusion as one is working on the project is to ensure that insights gathered from analysis of different types of data are in sync with each other. For example, in one of the projects company provided data from the shipping department to measure historical on-time delivery performance for selected products. Analysis of the shipment record showed exceptionally high delivery performance numbers which were at odds with other information available about the product's on-time delivery performance. On subsequent discussion, it was realized that the data shipping department was using to measure its delivery performance was using the modified date that planners had entered into the system after getting feedback from manufacturing that they won't be able to complete the job on time. On the other hand, temporary order surges sometimes drive the mentality to manufacture items in bigger batches, but data analysis will show the exact nature of the demand trends.

Several times during the projects, it is possible to get into the pitfalls of accepting rough estimates for the analysis. In a short-length interview, the response of a person is often affected by his/her emotional thinking. The information they provide is based on their specific point of view and experiences at that point in time. A better approach is to ask the interviewee for specific data which supports his/her statement. An effective interview is often based on a thorough discussion of the data.

It is also important to realize the significance of getting information from different sources to understand the problem at different levels. Talking with managers is important if one wants to understand the entire group of problems in a larger context. But in several cases, shop floor employees may give hints which help in identifying the root causes of a problem. Many ideas in projects came from discussions with the operators who were working on the identified group of products.

Lastly, it is also important to maintain a high-level view during any project. Responding with a specific recommendation to an issue is one thing and grouping all the recommendations together and making them work effectively is another.

Without maintaining a high-level approach, improvement ideas in one area may have a negative effect on the overall system performance. Working on all the areas simultaneously and keeping an eye on their interactions can help make the project recommendations more effective.

The cases discussed in this book clearly show how lean has played an important role in overcoming several challenges faced by managers of manufacturing organizations operating in a developed world. Over the years, a number of studies across the globe have highlighted the importance of lean methods and tools in improving operations and processes of manufacturing organizations. The power of lean tools is not hidden as success stories outsmart failures, and cases discussed earlier provide good supporting evidence. The success of the lean approach is because of its high effectiveness by reducing complexity and avoiding non-value-creating process steps. Lean, however, demands consistent and conscious efforts from the manufacturing organizations and those failing to meet these efforts are often at the other end of the success ladder. The implementation of the lean often requires manufacturing organizations to overcome several hindrances. As the world transition toward the new digital world (often referred to as the fourth industrial revolution or Industry 4.0), there is a greater need to explore how these disruptive technologies can aid the lean approach and help manufacturing organizations overcome the existing hindrances. Industry 4.0 is transforming the manufacturing sector by making factories smarter by creating a smart network of machines, products, components, properties, individuals, and ICT systems in the entire value chain. The case studies discussed here already show a good indication of how some of the Industry 4.0 technologies are assisting managers in lean implementation. Recent studies show that Industry 4.0 offers manufacturing organizations an estimated benefit by stabilizing lean processes with Industry 4.0 applications. As we are in the early phase of this digital transformation that is being mostly led by the developed world, there is a potential for manufacturing organizations of developed economies to harness the benefits of integrating Industry 4.0 technologies with a lean approach to maintain their competitiveness.

Bibliography

Aghayev, H., Garza-Reyes, J. A., Nadeem, S. P., Kumar, A., Kumar, V., Rocha-Lona, L., & González-Aleu, F. (2020). Lean readiness level of the Azerbaijan construction industry. In *Proceedings of the International Conference on Industrial Engineering and Operations Management*, Dubai, UAE, March, 10–12.

Belekoukias, I., Garza-Reyes, J. A., & Kumar, V. (2014). The impact of lean methods and tools on the operational performance of manufacturing organisations. *International Journal of Production Research, 52*(18), 5346–5366.

Garza-Reyes, J. A., Ates, E. M., & Kumar, V. (2015). Measuring lean readiness through the understanding of quality practices in the Turkish automotive suppliers industry. *International Journal of Productivity and Performance Management, 64*(8), 1092–1112.

Garza-Reyes, J. A., Betsis, I. E., Kumar, V., & Radwan Al-Shboul, M. A. (2018). Lean readiness – the case of the European pharmaceutical manufacturing industry. *International Journal of Productivity and Performance Management, 67*(1), 20–44.

Kolberg, D., & Zühlke, D. (2015). Lean automation enabled by industry 4.0 technologies. *IFAC-PapersOnLine, 48*(3), 1870–1875.

Liker, J. (2004). *The Toyota Way: 14 Management Principles from the World's Greatest Manufacturer*. McGraw-Hill Education ISBN-13:978-0071392310.

Möldner, A. K., Garza-Reyes, J. A., & Kumar, V. (2020). Exploring lean manufacturing practices' influence on process innovation performance. *Journal of Business Research, 106*, 233–249.

Mrugalska, B., & Wyrwicka, M. K. (2017). Towards lean production in industry 4.0. *Procedia Engineering, 182*, 466–473.

Saboo, A., Garza-Reyes, J. A., Er, A., & Kumar, V. (2014). A VSM improvement-based approach for lean operations in an Indian manufacturing SME. *International Journal of Lean Enterprise Research, 1*(1), 41–58.

Sanders, A., Elangeswaran, C., & Wulfsberg, J. P. (2016). Industry 4.0 implies lean manufacturing: Research activities in industry 4.0 function as enablers for lean manufacturing. *Journal of Industrial Engineering and Management (JIEM), 9*(3), 811–833.

Shah, R., & Ward, P. T. (2003). Lean manufacturing: Context, practice bundles, and performance. *Journal of Operations Management, 21*(2), 129–149.

Spearman, M. L., & Hopp, W. J. (2011). *Facory Physics* 3rd Edition. Waveland Pr Inc ISBN 13:978-1-57766-739-1.

Suri, R. (1998). *Quick Response Manufacturing: A Companywide Approach to Reducing Lead Times*. Productivity Press ISBN 9781563272011.

Tripathi, A. K., Tiwari, M. K., & Chan, F. T. S. (2005). Multi-agent-based approach to solve part selection and task allocation problem in flexible manufacturing systems. *International Journal of Production Research, 43*(7), 1313–1335.

Wagner, T., Herrmann, C., & Thiede, S. (2017). Industry 4.0 impacts on lean production systems. *Procedia CIRP, 63*, 125–131.

Yadav, G., Kumar, A., Luthra, S., Garza-Reyes, J. A., Kumar, V., & Batista, L. (2020). A framework to achieve sustainability in manufacturing organisations of developing economies using industry 4.0 technologies' enablers. *Computers in Industry*, 122, 103280.

Appendix

APPENDIX A: SAMPLE TRAVELER SHEET

Traveler Sheet

Traveler Sheet #	5
Part #	723155-008
Start Date	02/11/2016
Quantity	800
Schedule Ship Date	02/20/2016

Router Operation Step	Operation Description	Qty	Workstation #	Set-up Time Per Lot	IN Date	IN Time		OUT Date	OUT Time		Comment
100	Shear	800	25	1 hr.	2/11	11:30	AM	2/14	12:42	PM	
200	Form + Blank	800	9	30 min	2/14	1:36	PM	2/15	8:12	AM	
300	Pierce + Trim	400	2	45 min	2/15	1:17	PM	2/15	5:45	PM	
300	Pierce + Trim	400	2	45 min	2/16	11:13	AM	2/17	9:11	AM	
400	Paint	800	—	—	2/17	11:45	AM	2/19	4:45	PM	
							AM/PM			AM/PM	
							AM/PM			AM/PM	

Date Completed	2/19/2016
Time Completed	4:45 PM
Quantity Completed	800
Date Shipped	2/20/2016
Time Shipped	9:00 AM
Quantity Shipped	575

Started Pierce + Trim @ 1:17 pm on 2/15/2016
Finished Pierce + Trim @ 9:11 AM on 2/17/2016

Went to E coat on 2/17/2016 – 800 pcs.
Came back from E coat on 2/19/2016 – 800 pcs.

APPENDIX B: LABOR AND EQUIPMENT DETAILS USED IN MODEL

Details of the Labor Used in the Simulation Model of Current State

Operators	First Shift	Second Shift
Laser operator	2.0	0
Press brake operator	0.5	0
Stamping operator	0.7	0
Welding operator	1.5	0

Details of Equipment Used in the Simulation Model in Current State

Equipment	# In Group	% Time Available
135T Press brake	1	150% (1.5 Shifts)
260 Turret punch	1	100% (1 Shift)
2600 W Laser	1	200% (2 Shifts)
35T Press	1	65% (1 Shift)
4000 W Laser	1	200% (2 Shifts)
75KVA Spot welder	1	65% (1 Shift)
MIG Welder	1	50% (1 Shift)
Spot welder	1	50% (1 Shift)

APPENDIX C: LABOR AND EQUIPMENT DETAILS FOR PROPOSED CHANGES

Details of the Labor Used in the Simulation Model of Proposed State

Operators	First Shift	Second Shift
Laser operator	1.0	1
Press brake operator	0.4	0
Welding operator	1.0	0

Details of Equipment Used in the Simulation Model of Proposed State

Equipment	# In Group	% Time Available
135T Press brake	1	100% (1 Shift)
260 Turret punch	1	100% (1 Shift)
2600 W Laser	1	200% (2 Shifts)
35T Press	1	100% (1 Shift)
4000 W Laser	1	200% (2 Shifts)
75KVA Spot welder	1	100% (1 Shift)
MIG Welder	1	100% (1 Shift)
Spot Welder	1	100% (1 Shift)

APPENDIX D: CAUSE-AND-EFFECT DIAGRAM

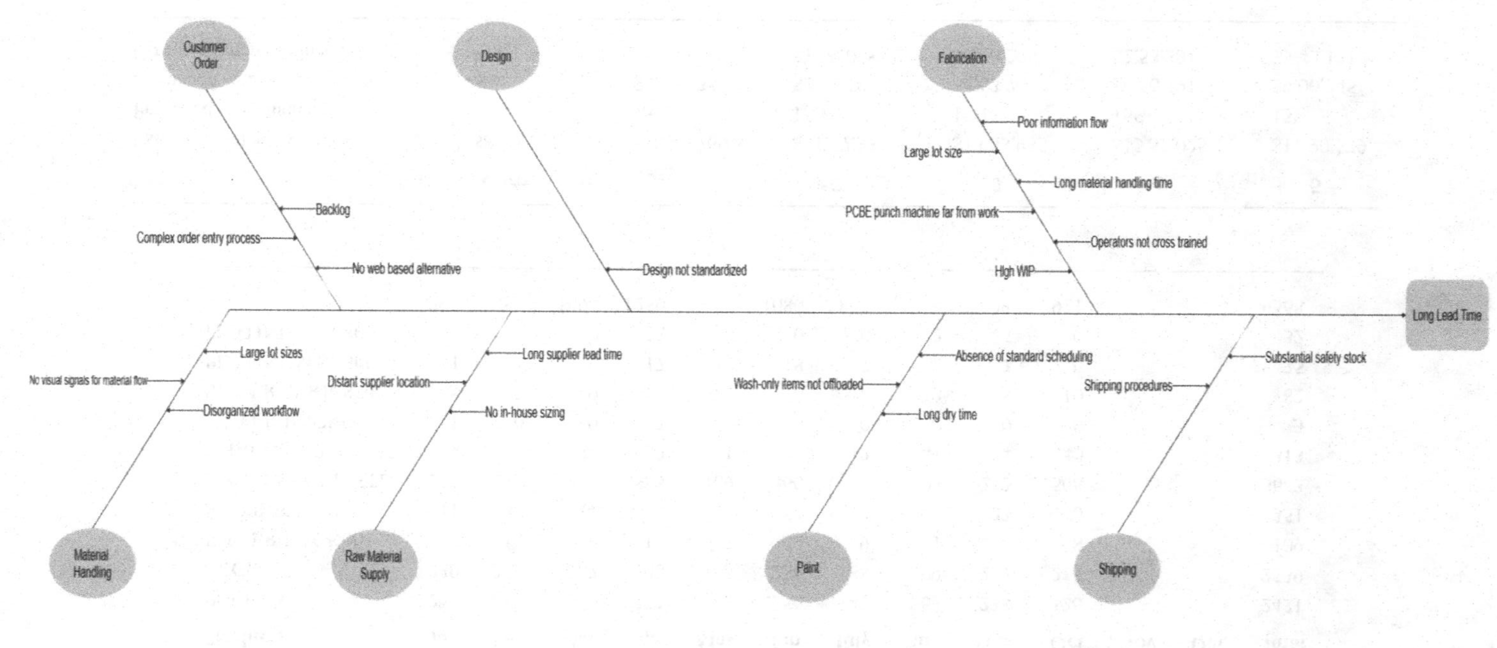

APPENDIX E: SALES GROWTH ESTIMATE

Products	Jan	Feb	Mar	Apr	May	Jun	July	Aug	Sept	Oct	Nov	Dec	Total
PCBE	255	282	306	272	282	256	141	263	238	126			2421
CTCR	239	272	267	208	305	254	188	309	282	215			2539
POWER RACK001	10	11	9	15	7	22	9	11	7	8			109
RED RACK001	11	33	14	69	46	54	14	71	23	16			351
CHALLENGER-STD	256	276	354	583	189	460	464	303	252	506			3643
CHALLENGER-SMC	18	2	126	59	1	10	12	41	4	40			313
POWER RACK002	1	0	0	2	0	0	0	0	0	0			3
SILVER SERIES001	5	7	9	12	14	6	2	12	5	10			82
PLATINUM SR500	1	2	9	17	12	18	7	2	7	0			75
PLATINUM SR800	2	1	0	22	0	0	0	2	5	0			32
Total	798	886	1094	1259	856	1080	837	1014	823	921			9568

Year	Current	1	2	3	4	5
Estimated sales revenue	$8,90,234	$10,23,769	$11,77,334	$13,53,935	$15,57,025	$17,90,579
Projected sales increase		15%	15%	15%	15%	15%
Estimated sales revenue increase	$0	$1,33,535	$2,87,100	$4,63,701	$6,66,791	$9,00,345
Estimated increase in profit		$50,743	$1,09,098	$1,76,206	$2,53,381	$3,42,131

APPENDIX F: ONE-TIME INVENTORY INVESTMENT SAVINGS

Product	Description	Qty. Per Product	Raw Material Savings (Units)	Raw Material Savings ($)	Punch WIP savings (Units)	Punch WIP Savings ($)	Machined WIP Savings (Units)	Machined WIP Savings ($)	Finished Goods Savings (Units)	Finished Goods Savings ($)
PCBE	Top	2	806	1153	830	1187	501	716	15	10,526
PCBE	Side	2	106	1234	–26	–303	96	1117		
PCBE	Base	2	204	1620	1213	9631	366	2906		

Product	Description	Qty. Per Product	Raw Material Savings (Units)	Raw Material Savings ($)	Punch WIP Savings (Units)	Punch WIP Savings ($)	Welded WIP Savings (Units)	Welded WIP Savings ($)	Finished Goods Savings (Units)	Finished Goods Savings ($)
CTCR	Side	2	1388	6329	0	0	5	25	35	172
CTCR	Cross	11	8549	3078	3879	1396				

APPENDIX G: TOTAL COST-BENEFIT SCHEDULE

Planning Horizon		Year 0	Year 1	Year 2	Year 3	Year 4	Year 5
Cash outflows:							
Depreciable							
Equipment		($60,000)					
Tooling		($6,860)	$0	$0	$0	$0	$0
Equipment relocation		($10,000)					
Non-Depreciable							
RPA		($1,05,000)	($15,000)	($15,000)	($15,000)	($15,000)	($15,000)
Direct labor		($36,000)	($36,000)	($36,000)	($36,000)	($36,000)	($36,000)
Raw material change		$0	($13,452)	($16,815)	($21,019)	($26,274)	($32,842)
	Total	($2,17,860)	($64,452)	($67,815)	($72,019)	($77,274)	($83,842)
Cash inflows:							
Operational Cost Savings							
RPA			$75,000	$75,000	$75,000	$75,000	$75,000
Raw materials salvage			$2574	$3217	$4022	$5027	$6284
Inventory investment reduction		$0	$45,788	$0	$0	$0	$0
Savings in inventory holding cost		$0	$4,579	$4,579	$4,579	$4,579	$4,579
Additional Income							
Income from increased sales		$0	$50,743	$1,09,098	$1,76,206	$2,53,381	$3,42,131
	Total	$0	$1,78,684	$1,91,894	$2,59,807	$3,37,987	$4,27,994

(Continued)

Net cash flows:						
Net cash flows	($2,17,860)	$1,14,232	$1,24,079	$1,87,788	$2,60,713	$3,44,152
Net accumulation (years 0–5)	($2,17,860)	($1,03,628)	$20,451	$2,08,239	$4,68,952	$8,13,103
Net accumulation NPV(years 0–5)	($2,17,860)	($1,03,648)	($10,425)	$1,17,836	$2,79,718	$4,73,983
Net cash flows after tax						
Depreciation (straight-line)	$0	$12,000	$12,000	$12,000	$12,000	$12,000
Taxable income stream	$0	$1,02,232	$1,12,079	$1,75,788	$2,48,713	$3,32,152
Taxes paid	$0	$21,469	$23,537	$36,915	$52,230	$69,752
Net cash flows	($2,17,860)	$92,763	$1,00,543	$1,50,872	$2,08,483	$2,74,400
Net accumulation after tax (years 0–5)	($2,17,860)	($1,25,097)	($24,554)	$1,26,318	$3,34,801	$6,09,201
Net accumulation NPV (years 0–5)	($2,17,860)	($1,21,391)	($45,852)	$57,196	$1,86,648	$3,41,539
Payback period:	2.30					
Net present value:	$341,539					
Internal rate of return:	53%					

APPENDIX H: PRODUCTS IN EACH CONTRACT MANUFACTURING GROUP

The custom manufacturing product set consists of three different product categories, namely Group 100, 200, and 300. The products within these categories are shown below:

Product Groups	Product Models	Annual Revenue	Units Sold
Group 100	L-brand	$7.8M	2610
	D-Brand		3600
	CMC		1710
	Creation cabinets		360
	Slimline		180
	Prototype products		180
	Preston products		90
	Greyson		90
	Flexer		90
	Modern designs		90
Group 200	HD small batch	$9M	6150
	Vector		5100
	Chester		3750
Group 300	Big store HVLM	$13.2M	20,000
	Spare parts		15,000

APPENDIX I: WEEKLY DEMAND OF CMC CONFIGURATION

The CMC unit is sold in four different configurations. The weekly demand for each configuration is shown below:

Week	Total	Config 1	Config 2	Config 3	Config 4
1	6	3	0	0	3
2	6	1	1	0	4
3	14	12	0	0	2
4	9	1	8	0	0
5	11	6	4	1	0
6	9	0	5	1	3
7	14	8	4	0	2
8	6	3	0	0	3
9	13	5	0	4	4
10	4	2	0	0	2
11	6	6	0	0	0
12	6	1	3	0	2
13	9	8	1	0	0
14	7	6	0	0	1
15	13	5	1	6	1
16	15	0	7	8	0
17	10	4	5	0	1
18	20	7	2	3	8
19	6	5	1	0	0
20	14	8	3	0	3
21	7	5	0	2	0
22	6	0	0	0	6
23	2	1	1	0	0
24	10	10	0	0	0
25	3	0	1	2	0
26	2	2	0	0	0
27	6	5	1	0	0
28	6	0	0	1	5
29	15	2	1	8	4
30	9	8	0	1	0
31	10	7	2	1	0
32	5	1	2	1	1
33	6	1	1	0	4
34	9	2	2	4	1
35	9	3	2	0	4
36	13	4	2	1	6
37	2	0	0	1	1
38	6	5	0	1	0
39	5	2	0	3	0
40	5	2	2	1	0
41	14	13	1	0	0
42	7	6	0	1	0
43	8	2	3	3	0
44	3	2	0	1	0
45	12	3	0	8	1
46	9	2	0	3	4
47	7	2	4	0	1
48	13	3	10	0	0

APPENDIX J: WEEKLY DEMAND OF CMC COMPONENTS

Different configurations of CMC are made with components A, B, C, D, E, F, G, H, and I. Weekly demand for all of these components for 48 weeks is shown below:

Week	Part A	Part B	Part C	Part D	Part E	Part F	Part G	Part H	Part I
1	150	6	6	21	24	24	30	36	18
2	132	6	6	23	22	16	24	36	14
3	400	14	14	44	56	76	80	84	52
4	206	9	9	35	20	22	22	54	20
5	286	11	11	36	34	50	48	64	36
6	188	9	9	34	24	22	26	52	20
7	368	14	14	48	48	60	64	84	44
8	150	6	6	21	24	24	30	36	18
9	302	13	13	39	44	62	62	70	44
10	100	4	4	14	16	16	20	24	12
11	180	6	6	18	24	36	36	36	24
12	136	6	6	23	18	16	20	36	14
13	262	9	9	28	34	50	50	54	34
14	200	7	7	22	28	38	40	42	26
15	300	13	13	35	38	70	60	66	48
16	298	15	15	44	30	62	46	74	46
17	250	10	10	36	30	36	38	60	28
18	468	20	20	67	70	80	90	114	60
19	172	6	6	19	22	32	32	36	22
20	366	14	14	48	50	60	66	84	44
21	186	7	7	19	24	42	38	38	28
22	120	6	6	24	24	12	24	36	12
23	52	2	2	7	6	8	8	12	6
24	300	10	10	30	40	60	60	60	40
25	58	3	3	8	6	14	10	14	10
26	60	2	2	6	8	12	12	12	8
27	172	6	6	19	22	32	32	36	22
28	118	6	6	22	22	16	24	34	14
29	306	15	15	42	42	70	62	74	50
30	258	9	9	26	34	54	52	52	36
31	272	10	10	31	34	52	50	58	36
32	112	5	5	17	14	18	18	28	14
33	132	6	6	23	22	16	24	36	14
34	196	9	9	26	24	42	36	46	30
35	214	9	9	33	32	30	38	54	24
36	302	13	13	46	46	46	56	76	36
37	38	2	2	6	6	8	8	10	6
38	168	6	6	17	22	36	34	34	24
39	114	5	5	12	14	30	24	24	20
40	122	5	5	16	14	22	20	28	16
41	412	14	14	43	54	80	80	84	54
42	198	7	7	20	26	42	40	40	28
43	180	8	8	24	20	36	30	42	26
44	78	3	3	8	10	18	16	16	12
45	254	12	12	29	32	68	54	56	46
46	194	9	9	28	30	38	40	48	28
47	168	7	7	26	20	22	24	42	18
48	310	13	13	49	32	38	38	78	32

Index

Printed in the United States
by Baker & Taylor Publisher Services

Printed in the United States
by Baker & Taylor Publisher Services